IDENTITIES
Fundamental Identities

1. $\sec \theta = \dfrac{1}{\cos \theta}$

2. $\csc \theta = \dfrac{1}{\sin \theta}$

3. $\cot \theta = \dfrac{1}{\tan \theta}$

4. $\tan \theta = \dfrac{\sin \theta}{\cos \theta}$

5. $\cot \theta = \dfrac{\cos \theta}{\sin \theta}$

6. $\cos^2 \theta + \sin^2 \theta = 1$

7. $\tan^2 \theta + 1 = \sec^2 \theta$

8. $1 + \cot^2 \theta = \csc^2 \theta$

Sum and Difference Identities

9. $\cos(\alpha + \beta) = \cos \alpha \cos \beta - \sin \alpha \sin \beta$
   $\cos(\alpha - \beta) = \cos \alpha \cos \beta + \sin \alpha \sin \beta$

10. $\sin(\alpha + \beta) = \sin \alpha \cos \beta + \cos \alpha \sin \beta$
    $\sin(\alpha - \beta) = \sin \alpha \cos \beta - \cos \alpha \sin \beta$

11. $\tan(\alpha + \beta) = \dfrac{\tan \alpha + \tan \beta}{1 - \tan \alpha \tan \beta}$

    $\tan(\alpha - \beta) = \dfrac{\tan \alpha - \tan \beta}{1 + \tan \alpha \tan \beta}$

Double-Angle Identities

12. $\cos 2\theta = \cos^2 \theta - \sin^2 \theta$
    $= 2\cos^2 \theta - 1$
    $= 1 - 2\sin^2 \theta$

13. $\sin 2\theta = 2 \sin \theta \cos \theta$

14. $\tan 2\theta = \dfrac{2 \tan \theta}{1 - \tan^2 \theta}$

Half-Angle Identities

15. $\cos \tfrac{1}{2}\theta = \pm\sqrt{\dfrac{1 + \cos \theta}{2}}$

16. $\sin \tfrac{1}{2}\theta = \pm\sqrt{\dfrac{1 - \cos \theta}{2}}$

17. $\tan \tfrac{1}{2}\theta = \dfrac{1 - \cos \theta}{\sin \theta}$

    $= \dfrac{\sin \theta}{1 + \cos \theta}$

Product Identities

18. $2 \cos \alpha \cos \beta = \cos(\alpha - \beta) + \cos(\alpha + \beta)$

19. $2 \sin \alpha \sin \beta = \cos(\alpha - \beta) - \cos(\alpha + \beta)$

20. $2 \sin \alpha \cos \beta = \sin(\alpha + \beta) + \sin(\alpha - \beta)$

Sum Identities

21. $\cos \alpha + \cos \beta = 2 \cos\left(\dfrac{\alpha + \beta}{2}\right)\cos\left(\dfrac{\alpha - \beta}{2}\right)$

22. $\cos \alpha - \cos \beta = -2 \sin\left(\dfrac{\alpha + \beta}{2}\right)\sin\left(\dfrac{\alpha - \beta}{2}\right)$

23. $\sin \alpha + \sin \beta = 2 \sin\left(\dfrac{\alpha + \beta}{2}\right)\cos\left(\dfrac{\alpha - \beta}{2}\right)$

24. $\sin \alpha - \sin \beta = 2 \sin\left(\dfrac{\alpha - \beta}{2}\right)\cos\left(\dfrac{\alpha + \beta}{2}\right)$

Miscellaneous Identities

25. $\cos(\pi/2 - \theta) = \sin \theta$

26. $\sin(\pi/2 - \theta) = \cos \theta$

27. $\tan(\pi/2 - \theta) = \cot \theta$

28. $\cos(-\theta) = \cos \theta$

29. $\sin(-\theta) = -\sin \theta$

30. $\tan(-\theta) = -\tan \theta$

LAW OF COSINES

$a^2 = b^2 + c^2 - 2bc \cos \alpha$
$b^2 = a^2 + c^2 - 2ac \cos \beta$
$c^2 = a^2 + b^2 - 2ab \cos \gamma$

$\cos \alpha = \dfrac{b^2 + c^2 - a^2}{2bc}$

$\cos \beta = \dfrac{a^2 + c^2 - b^2}{2ac}$

$\cos \gamma = \dfrac{a^2 + b^2 - c^2}{2ab}$

LAW OF SINES

$\dfrac{\sin \alpha}{a} = \dfrac{\sin \beta}{b} = \dfrac{\sin \gamma}{c}$

TRIGONOMETRY
FOR COLLEGE STUDENTS

TRIGONOMETRY
FOR COLLEGE STUDENTS
Karl J. Smith
Santa Rosa Junior College

BROOKS/COLE PUBLISHING COMPANY
MONTEREY, CALIFORNIA
A Division of Wadsworth Publishing Company, Inc.

ISBN: 0-8185-0198-7
L.C. Catalog Card No.: 76-19454
Printed in the United States of America

10 9 8 7 6 5 4 3

Manuscript Editor: *Micky Lawler*
Production Editor: *Cece Munson*
Interior and Cover Design: *John Edeen*
Illustrations: *John Foster*
Cover Art: *Tecumseh, ©1973, by Norton Starr*

To my mother,
Rosamond,
with love and thanks.

CREDITS
This page constitutes an extension of the copyright page.

PREFACE

Trigonometry for College Students introduces the basic concepts of trigonometry as well as several related applications. The goal of this textbook is to present students with the skills necessary to continue their studies in mathematics. Because of the great availability of inexpensive electronic calculators, I have emphasized the nature of trigonometric functions rather than the computational aspects of the subject.

The style of the text is informal, and I have written the book with the student always in mind. There are more than 300 completely solved examples, and the more than 1200 problems are graded in difficulty. Answers to most odd-numbered problems appear in the back of the book. The easiest problems are the "A Problems," and the more difficult ones are labeled "B Problems." Most problem sets also include a section of "Mind Bogglers," which deal with such diverse ideas as chain letters, biorhythm charts, and Euler's Formula.

Functions and graphs are presented in Chapter 1. Although much of the material in this chapter is review, it is worthwhile to consider it carefully, since it lays the groundwork for the course. Instead of graphing the parabola $f(x) = ax^2 + bx + c$ in the usual algebraic manner, we learn in Chaper 1 to graph it as the sum of a standard-position parabola $f_1(x) = ax^2$ and a line $f_2(x) = bx + c$ so that $f(x) = f_1(x) + f_2(x)$. The groundwork for the inverse trigonometric functions is also set in Chapter 1.

Chapter 2 provides definitions and graphs of the trigonometric functions and therefore serves as an introduction to trigonometry that sets the stage for further study in mathematics. Chapter 3 presents a detailed study of trigonometric identities. However, since Chapters 3 and 4 are independent, the right-triangle trigonometry of Chapter 4 could be studied prior to Chapter 3 if the instructor desires.

Chapter 5 considers the solution of oblique triangles. It is assumed that most students have access to an electronic calculator. For those who do not, logarithms are presented in Appendix B. I would suggest that, without the use of calculators, accuracy requirements be relaxed.

Most students will have had only a brief introduction to complex numbers, so a rather complete development of the topic, including De Moivre's Formula, is presented in Chapter 6. The chapter concludes with a discussion of the polar coordinate system and of the graphing of polar curves.

I am grateful to the following persons who read the manuscript and offered helpful suggestions: Paul W. Haggard of East Carolina University, Franz X. Hiergeist of West Virginia University, Mustafa Munen of Macomb County Community College, Carla B. Oviatt of Montgomery College, and Robert J. Wisner of New Mexico State University.

Karl J. Smith

CONTENTS

1

FUNCTIONS AND GRAPHS   2

2

TRIGONOMETRIC FUNCTIONS   60

ix

6

COMPLEX NUMBERS   202

TRIGONOMETRY
FOR COLLEGE STUDENTS

Trigonometry is a branch of mathematics that deals with the properties and applications of six functions: the sine, cosine, tangent, cotangent, secant, and cosecant. To understand these trigonometric functions, we must first understand the idea of a function and its graph. Chapter 1 reviews this topic and introduces the notation and background needed for a study of trigonometry.

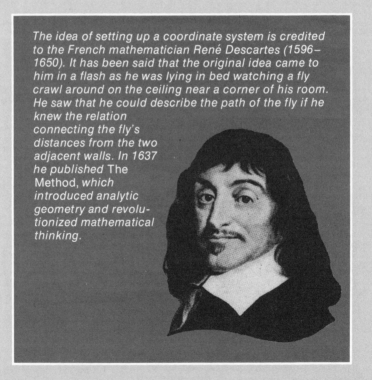

The idea of setting up a coordinate system is credited to the French mathematician René Descartes (1596–1650). It has been said that the original idea came to him in a flash as he was lying in bed watching a fly crawl around on the ceiling near a corner of his room. He saw that he could describe the path of the fly if he knew the relation connecting the fly's distances from the two adjacent walls. In 1637 he published The Method, *which introduced analytic geometry and revolutionized mathematical thinking.*

1

FUNCTIONS AND GRAPHS

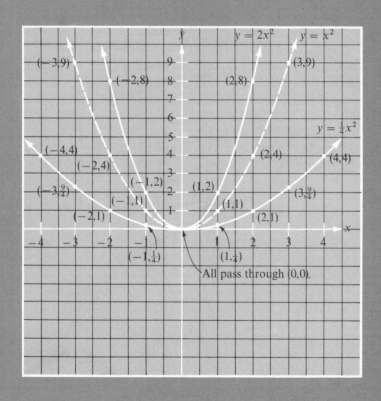

1.1 RELATIONS AND FUNCTIONS (*Feature the Cone and the Function Thereof*)

In mathematics we are frequently concerned with the relationships between various sets of numbers. For example, suppose we have an ice-cream container shaped like a right circular cone, as shown.

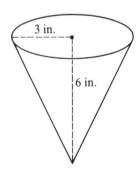

Suppose also that Clumsy Carp in the *B.C.* cartoon wishes to experiment by filling this cone with sand. He finds that the radius of sand in the cone is always $\frac{1}{2}$ the height of the sand and that the volume is found by

$$V = \frac{1}{12}\pi h^3.$$

He can construct a table of values showing the height and the volume for certain values.

Height	1	2	3	4	5	6
Volume	$\dfrac{\pi}{12}$	$\dfrac{2}{3}\pi$	$\dfrac{9}{4}\pi$	$\dfrac{16}{3}\pi$	$\dfrac{125}{12}\pi$	18π

This table can be written as a set of *ordered pairs:* $\{(1, \pi/12), (2, \frac{2}{3}\pi), (3, \frac{9}{4}\pi), \ldots\}$.

A set of ordered pairs is called a *relation.* In our example we are considering ordered pairs (h, V) such that $V = \frac{1}{12}\pi h^3$. The variable associated with the first component is called the *independent variable*, and the variable associated with the second component is called the *dependent variable.*

The set of possible replacements for the independent variable is sometimes limited to a certain set of numbers. In our cartoon example, h (the height of the sand in the cone) cannot be negative (why?) and cannot be larger than 6 inches, since the conical container has this height. We say that the *domain* is the set of possible replacements for the independent variable. Thus the domain, D, is

$$D = \{h \mid 0 < h \le 6\}.^*$$

The set of replacements for the dependent variable is the *range* of the relation.

It is often useful to draw a graph of a relation. To do so, we set up a *rectangular coordinate system*, as shown in Figure 1.1. A rectangular coordinate system is sometimes called a *Cartesian coordinate system* in honor of René Descartes.

The horizontal number line is called the *x-axis* (sometimes called the axis of abscissas), and *x* represents the first component of the ordered pair; the vertical number line is called the *y-axis* (sometimes called the axis of ordinates), and *y* represents the second component of the ordered pair. The point of intersection of the axes is the *origin.* These number lines divide the plane into four regions, called *quadrants*, as shown in Figure 1.1.

When plotting ordered pairs called points, or drawing a graph, it is important to label the axes and indicate the scale. Although it is often convenient to choose the same scale or unit for both axes, it is not necessary to do so. If we wish to plot the points obtained from the cartoon example

$$\left\{\left(1, \frac{\pi}{12}\right), \left(2, \frac{2}{3}\pi\right), \left(3, \frac{9}{4}\pi\right)\right\}$$

---

*This is the so-called set-builder notation. It is read "*the set of all h such that* zero is less than *h* and *h* is less than or equal to six."

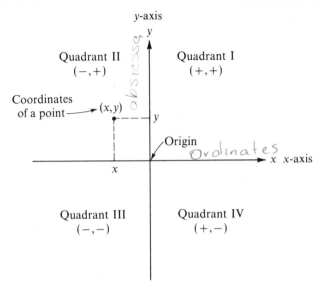

FIGURE 1.1   Cartesian coordinate system

and choose the same scale for both axes, we would need to approximate π and calculate values for the second components. However, if we choose the scale shown in Figure 1.2(a), the second components can be plotted directly. Figure 1.2(b) shows the graph of the relation for all values of the domain. We also see that the range, R, is given by

$$R = \{V \mid 0 < V \leq 18\pi\}.$$

In this book we will limit our study to a special type of relation called a *function*.

Definition of a Function: A **function** is a relation for which each member of the domain is associated with exactly one member of the range.

EXAMPLES:

1. $\{(0, 0), (1, 1), (2, 2), (3, 3), (4, 4)\}$
     $D = \{0, 1, 2, 3, 4\}$; $R = \{0, 1, 2, 3, 4\}$

D	R
0 $\longrightarrow$	0
1 $\longrightarrow$	1
2 $\longrightarrow$	2
3 $\longrightarrow$	3
4 $\longrightarrow$	4

This relation is a function.

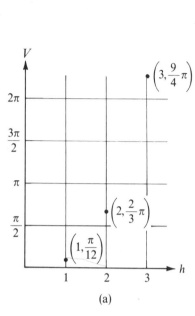

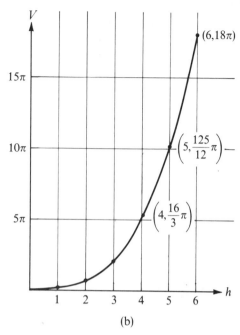

(a)   (b)

FIGURE 1.2

2. $\{(0, 0), (1, 1), (-1, 1), (2, 4), (-2, 4)\}$
    $D = \{0, 1, -1, 2, -2\}; R = \{0, 1, 4\}$

D	R
0 ⟶	0
1 ⟩	
-1 ⟩	1
2 ⟩	
-2 ⟩	4

This relation is a function. Notice that each *first* component is associated with exactly one second component.

3. $\{(0, 0), (1, 1), (1, -1), (4, 2), (4, -2)\}$
    $D = \{0, 1, 4\}; R = \{0, 1, -1, 2, -2\}$

D	R
0 ⟶	0
1 ⟨	1
	-1
4 ⟨	2
	-2

This relation is *not* a function, since the first component, 1, is associated with *both* 1 and −1. Notice that, to claim that a particular relation is not a function, we need find only one member of the domain that is associated with more than one member of the range.

4. $y = 2x + 3$
$D = \{\text{real numbers}\}; \ R = \{\text{real numbers}\}$
This equation specifies a relation that is the set of ordered pairs satisfying the equation; that is, when we write $y = 2x + 3$, we mean $\{(x, y) | y = 2x + 3\}$. This set is a function, since each member of the domain will yield exactly one member in the range.

If a relation is a function, we use a notational shorthand by assigning it a "name," which is generally in the form of some letter of the alphabet. For example, if the functions of Examples 2 and 4 are named by

$$f = \{(0, 0), (1, 1), (-1, 1), (2, 4), (-2, 4)\};$$
$$g = \{(x, y) | y = 2x + 3\},$$

we can distinguish between these functions by calling one f and the other g. Suppose we are given some member of the domain of each of these functions (say -2) and wish to find the second components associated with -2:

function f: if $x = -2$, then $y = 4$ from the ordered pair $(-2, 4)$;
function g: if $x = -2$, then $y = 2(-2) + 3$
$$= -1, \text{ giving us the ordered pair } (-2, -1).$$

These responses are rather lengthy, so we go one step further and use the symbol

$$f(x)$$

to denote the second component associated with x for the function f. The symbol $f(x)$ is read as "f of x." Using this new notation, we write

$$f(-2) = 4 \quad \text{and} \quad g(-2) = -1.$$

EXAMPLES: Let $f_1 = \{(95, 1), (4, 3), (3, 17)\}$;
$f_2 = \{(x, y) | y = 5x\}$, which we shorten by writing $f_2(x) = 5x$;
$f_3(x) = 3x^2 + 2$.

Thus:

This is a member of the domain (a first component).

1. $f_1(3) = 17$.

This is the member of the range (a second component) that is associated with 3.

2. $f_2(3) = 15$, since $5 \cdot 3 = 15$.

Replace x by 3 in $f_2(x) = 5x$.

3. $f_2(4) = 20$, since $5 \cdot 4 = 20$.

    Replace x by 4.

4. $f_3(4) = 3(4)^2 + 2$
    $= 50$.

PROBLEM SET 1.1

A Problems

Classify the sets of Problems 1–10 as relations, functions, both, or neither.

1. $\{(6, 3), (9, 4), (7, -1), (5, 4)\}$ Neither Both

2. $\{6, 9, 7, 5\}$ Neither

3. $\{(3, 6), (4, 9), (-1, 7), (4, 5)\}$ Relation

4. $\{(x, y) \mid y = 5x + 2\}$ Both

5. $\{(x, y) \mid y \leq 5x + 2\}$ Both Relation

6. $\{(x, y) \mid y = -1$ if x is a rational number and $y = 1$ if x is an irrational number$\}$

7. $\{(x, y) \mid y =$ closing price of IBM stock on January 2 of year $x\}$ FUNCTION

8. $\{(x, y) \mid x =$ closing price of Xerox stock on July 1 of year $y\}$ FUNCTION

9. $\{(x, y) \mid (x, y)$ is a point on a circle with center $(2, 3)$ and radius $4\}$ RELATION

10. $\{(x, y) \mid (x, y)$ is a point on a line passing through $(2, 3)$ and $(4, 5)\}$ FUNCTION

For Problems 11–14 use the following table, which reflects the purchasing power of the dollar from October 1944 to October 1974 (Source: U.S. Bureau of Labor Statistics, Consumer Division). *Let x represent the year; let the domain be the set $\{1944, 1954, 1964, 1974\}$; let*

$r(x)$ = price of a pound of round steak;
$s(x)$ = price of a 5-pound bag of sugar;
$b(x)$ = price of a loaf of bread;
$c(x)$ = price of a pound of coffee;
$e(x)$ = price of a dozen eggs;
$m(x)$ = price of a half-gallon of milk;
$g(x)$ = price of a gallon of gasoline.

Year	Round steak (1 lb.)	Sugar (5-lb. bag)	Bread (loaf)	Coffee (1 lb.)	Eggs (1 doz.)	Milk (half-gallon)	Gasoline (1 gallon)
1944	$0.45	$0.34	$0.09	$0.30	$0.64	$0.29	$0.21
1954	0.92	0.52	0.17	1.10	0.60	0.45	0.29
1964	1.07	0.59	0.21	0.82	0.57	0.48	0.30
1974	1.78	2.08	0.36	1.31	0.84	0.78	0.53

11. Find:
     a. $r(1964)$.   b. $m(1954)$.   c. $g(1944)$.   d. $c(1974)$.

12. Find:
     a. $s(1974) - s(1944)$.   b. $b(1974) - b(1944)$.

13. a. Find the change in the price of eggs from 1944 to 1974.
     b. Write the change in the price of eggs using functional notation.

14. a. Find:

$$\frac{g(1944 + 30) - g(1944)}{30}.$$

     b. In words, can you attach any meaning to the number found in (a)?

In Problems 15–20 let $f(x) = 3x^2$, $g(x) = 4x - 1$, $h(x) = 5 - 2x$, $k(x) = 500x$, and $V(r) = 2\pi r^3$.
Find and plot the ordered pairs specified for each problem.

15. a. $f(-1)$ b. $f(0)$ c. $f(1)$ d. $f(2)$
16. a. $g(-2)$ b. $g(0)$ c. $g(2)$ d. $g(4)$
17. a. $h(5)$ b. $h(6)$ c. $h(8)$ d. $h(10)$
18. a. $k(5)$ b. $k(6)$ c. $k(7)$ d. $k(8)$
19. a. $k(0.01)$ b. $k(0.05)$ c. $k(0.08)$ d. $k(0.1)$
20. a. $V(1)$ b. $V(2)$ c. $V(3)$ d. $V(4)$

B Problems

In Problems 21–26 compute the given value where $f(x) = x^2$ and $g(x) = 4x + 3$.

21. a. $f(w)$ b. $f(h)$ c. $f(w + h)$ d. $f(w) + f(h)$
22. a. $g(s)$ b. $g(t)$ c. $g(s + t)$ d. $g(s) + g(t)$
23. a. $f(x^2)$ b. $f(\sqrt{x})$ c. $f(x + h)$ d. $f(-x)$
24. a. $g(x^2)$ b. $g(\pi)$ c. $g(x + \pi)$ d. $g(-x)$

25. a. $f(x + h) - f(x)$ b. $\dfrac{f(x + h) - f(x)}{h}$

26. a. $g(x + h) - g(x)$ b. $\dfrac{g(x + h) - g(x)}{h}$

27. If $f(x) = x/(x - 1)(x + 3)$ and $g(x) = x/(x - 1)(x + 2)$, find:
     a. $f(3) + g(4)$.   b. $f(h) + g(h)$.

28. If $f(x) = \sqrt{x}$ and $g(x) = \sqrt{x + 1}$, find:
     a. $f(27) + g(11)$.   b. $g(a)/f(a)$.

29. If $f(x) = \sqrt{x^2 - x + 1}$ and $g(x) = \sqrt{x^3 + 1}$, find:
     a. $g(3)/f(3)$.   b. $g(b)/f(b)$.

30. If $f(x) = (x + 1)/(x^2 - 4)$ and $g(x) = x/(x^2 + x - 2)$, find:
     a. $f(3) - g(3)$.   b. $f(w) - g(w)$.

$$x^2 + 2x - 3$$
$$3$$
$$9 + 6 - 3$$

31. Use the table given for Problems 11–14.
   a. What was the average increase in the price of sugar per year from 1944 to 1954? Write this increase using functional notation.
   b. What was the average increase in the price of sugar per year from 1944 to 1964? Write this increase using functional notation.
   c. What was the average increase in the price of sugar per year from 1944 to 1974? Write this increase using functional notation.
   d. What was the average increase in the price of sugar per year from 1944 to 1944 + h, where h is an unspecified number of years?

32. Using the table given for Problems 11–14, answer the questions posed in Problem 31, but substitute the price of coffee for the price of sugar.

33. *Calculator Problem.* If an object is projected vertically from the ground at 112.43 feet per second (fps), then (neglecting air resistance) the distance d of the object above the ground at time t is given by the equation

$$d = 112.43t - 16t^2.$$

What is the height at $t = 1.4$ sec and at $t = 2.6$ sec? If the relation is specified by (t, d), what is the domain?

34. *Calculator Problem.* The projectile of Problem 33 has velocity v at time t given by

$$v = 112.43 - 32t.$$

What is the velocity for $t = 1.4$ sec and $t = 2.6$ sec? If the relation is specified by (t, v), what is the domain?

Mind Bogglers

35. a. Referring to the *Peanuts* cartoon, let x represent the number of mailings and y represent the total number of letters in that mailing. Assume that the six letters Charlie Brown is writing constitute the first mailing, and write the first five ordered pairs for this relation.
   b. Is the relation defined in (a) a function?
   c. Write an equation that defines the relation in (a).
   d. Why do you think it is illegal to mail chain letters? Be specific in your discussion.

(continued)

36. a. A frog is sitting at the bottom of a pit 50 feet deep. It climbs 30 feet a day but slips
back 20 feet each night. How long will it take the frog to get out of the pit?
b. Is the frog's position in the pit a function of time?
c. Suppose the frog starts in a pit P feet deep and climbs C feet a day while slipping
back N feet at night. Describe a procedure that will give D, the number of days to
freedom.

1.2 DISTANCE AND ANGLES *(What's It All About, α?)*

Functions and coordinate systems are essential to the study of trigonometry.
Distances and angles are also essential. Consider two points, $P_1(x_1, y_1)$ and
$P_2(x_2, y_2)$, and the segment connecting P_1 and P_2. From algebra the distance
between P_1 and P_2 is denoted by $|P_1 P_2|$ and is found by applying the Pythag-
orean Theorem to the triangle formed by P_1, P_2, and $Q(x_2, y_1)$. (See Figure 1.3.)

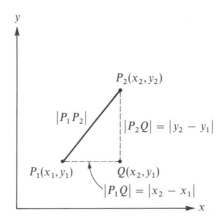

FIGURE 1.3   Distance between points

> *Distance Formula:* The **distance** between $P_1(x_1, y_1)$ and $P_2(x_2, y_2)$ is found by
>
> $$|P_1 P_2| = \sqrt{(x_2 - x_1)^2 + (y_2 - y_1)^2}.$$

EXAMPLES:

1. If $P_1(3, 2)$ and $P_2(9, 10)$, then $|P_1 P_2| = \sqrt{(9 - 3)^2 + (10 - 2)^2}$

$$= \sqrt{36 + 64}$$
$$= \sqrt{100}$$
$$= 10.$$

2. If $A(-5, 3)$ and $B(-2, -6)$, then $|AB| = \sqrt{[(-2) - (-5)]^2 + [(-6) - 3]^2}$

$$= \sqrt{9 + 81}$$
$$= \sqrt{90}$$
$$= 3\sqrt{10}.$$

Note that the distance between two points is always a nonnegative number; it is 0 only when the points coincide.

We can apply the distance formula to find the equation of a circle of radius r with center at the origin. If $P(x, y)$ represents any point on the circle, then the distance between P and O is r.

$$|PO| = r$$
$$\sqrt{(x - 0)^2 + (y - 0)^2} = r$$
$$\sqrt{x^2 + y^2} = r$$
$$x^2 + y^2 = r^2$$

We say that $x^2 + y^2 = r^2$ is the equation of the circle with center at the origin and radius r, as shown in Figure 1.4. In general, the equation of any graph sets up a one-to-one correspondence. *A graph of an equation* and *an equation of a graph* are closely connected: there is a one-to-one correspondence between the set of all ordered pairs (x, y) that satisfy the equation and the set of all ordered pairs (x, y) that lie on the curve.

If $r = 1$, we have the equation

$$x^2 + y^2 = 1,$$

whose graph is a circle centered at the origin with radius 1. This curve is important in trigonometry and is called the *unit circle*. Part of a circle is called an *arc*.

Circles can be related to angles. In geometry you probably defined an angle as the union of two rays with a common endpoint. In trigonometry, however, we use a more general definition.

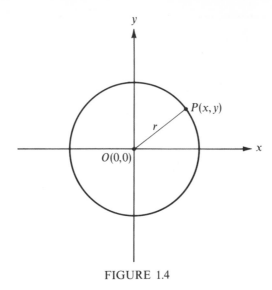

FIGURE 1.4

Definition of an Angle: An **angle** is formed by rotating a ray about its endpoint (called the *vertex*) from some initial position (called the *initial side*) to some terminal position (called the *terminal side*). The measure of an angle is the amount of rotation. An angle is also formed if a line segment is rotated about one of its endpoints.

If the rotation is in a counterclockwise direction, the measure of the angle is called *positive*. If the rotation is in a clockwise direction, the measure is called *negative*. We will use the notation $\angle ABC$ to denote the measure of an angle with vertex B and points A and C (different from B) on the sides. $\angle B$ will denote the measure of an angle with vertex at B, and we will use a curved arrow to denote the direction and amount of rotation. If no arrow is shown, the measure of the angle is considered to be the smallest positive rotation. We also use lower-case Greek letters to denote the measure of angles. Some examples are shown.

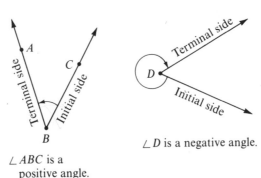

$\angle ABC$ is a positive angle.

$\angle D$ is a negative angle.

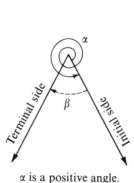

α is a positive angle.
β is a negative angle.

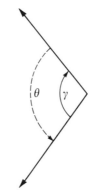

θ is a positive angle.
γ is a negative angle.

Table of Commonly Used Greek Letters*	
Symbol	*Name*
α	alpha
β	beta
γ	gamma
θ	theta
λ	lambda
ϕ or φ	phi
ω	omega

Examples of Angles

*π (pi) is a lowercase Greek letter that will not be used to represent an angle. It denotes an irrational number approximately equal to 3.141592654.

Many angles will have a Cartesian coordinate system superimposed so that the vertex is at the origin and the initial side is along the positive x-axis. In this case the angle is in *standard position*. Angles in standard position having the same terminal sides are *coterminal angles*. Given any angle α, there is an un-limited number of coterminal angles (some positive and some negative). In Figure 1.5, β is coterminal with α. Can you find other angles coterminal with α?

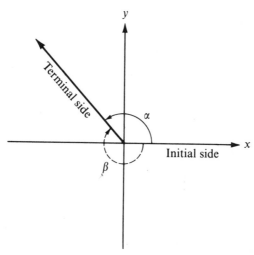

FIGURE 1.5   Standard-position angles α and β. α is a positive angle, and β is a negative angle. α and β are coterminal angles.

Let's consider a relationship among angles, circles, and the measures of angles. Let α be an angle in standard position with a point *P* not the vertex but on the terminal side. As this side is rotated through one revolution, the trace of the point *P* forms a circle. The measure of the angle is one revolution, but, since much of our work will be with amounts less than one revolution, we need to define measures of smaller angles. Historically, the most common scheme divides one revolution into 360 equal parts with each part called a *degree*. Sometimes even finer divisions are necessary, so a degree is divided into 60 equal parts, each called a *minute* (thus, $1° = 60'$). Furthermore, a minute can be divided into 60 equal parts, each called a *second* (thus, $1' = 60''$). For most applications we will write decimal parts of degrees instead of minutes and seconds. That is, 32.5° is preferred over 32° 30′.

In calculus and in scientific work, another measure for angles is defined. Let's draw a circle with any nonzero radius *r*. Next we measure out an arc whose length is *r*. Figure 1.6(a) shows the case in which $r = 1$ and Figure 1.6(b) shows $r = 2$. Regardless of our choice for *r*, the angle determined by this arc of length *r* is the same (it is labeled θ in Figure 1.6). This angle is used as a basic unit of measurement and is called a *radian*. Notice that the circumference generates an angle of one revolution. Since $C = 2\pi r$, and since the basic unit of measurement

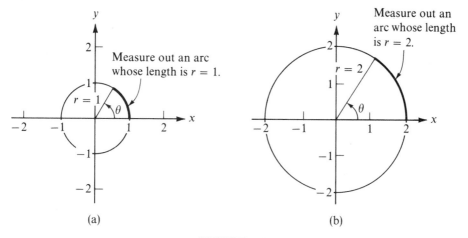

FIGURE 1.6

on the circle is r, we see that

$$\text{one revolution} = \frac{C}{r}$$

$$= \frac{2\pi r}{r}$$

$$= 2\pi.$$

Thus $\frac{1}{2}$ revolution is $\frac{1}{2}(2\pi) = \pi$ radians; $\frac{1}{4}$ revolution is $\frac{1}{4}(2\pi) = \pi/2$ radians.

Notice that, when we measure angles in radians, we are using *real numbers*. Because radian measure is used so frequently, when *no units* of measure for an angle are indicated, we agree that radian measure is understood.

EXAMPLES: Let $r = 1$. Draw the angles whose measures are given.

1. $\theta = 2$

2. $\theta = 3$

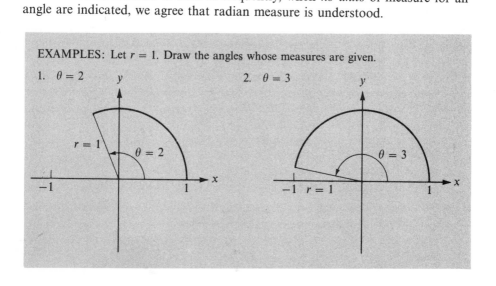

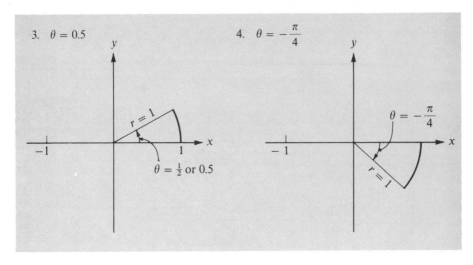

3. $\theta = 0.5$

$\theta = \frac{1}{2}$ or 0.5

4. $\theta = -\frac{\pi}{4}$

$\theta = -\frac{\pi}{4}$

We are led to the following relationships.

1 revolution is measured by $360°$;

1 revolution is measured by 2π.

If θ is the measure of any angle, then

$$\text{the number of revolutions} = \frac{\theta \text{ in degrees}}{360};$$

$$\text{the number of revolutions} = \frac{\theta \text{ in radians}}{2\pi}.$$

Therefore,

$$\frac{\theta \text{ in degrees}}{360} \quad \text{is equivalent to} \quad \frac{\theta \text{ in radians}}{2\pi}.$$

EXAMPLES:

1. Change $45°$ to radians:

$$\frac{45}{360} = \frac{\theta}{2\pi}$$

$$\frac{90\pi}{360} = \theta$$

$$\frac{\pi}{4} = \theta$$

An alternative method is to remember that π radians is 180°. That is, since 45° is $\frac{1}{4}$ of 180°, we know that the radian measure is $\pi/4$.

2. Change 1° to radians.

$$\frac{1}{360} = \frac{\theta}{2\pi}$$

$$\frac{\pi}{180} = \theta$$

Even though this solution is in the desired form, you might be interested in performing the above division on a calculator: $1° \approx 0.0174532925$.

3. Change 1 to degrees (remember that, if units are not specified, radians are understood).

$$\frac{\theta}{360} = \frac{1}{2\pi}$$

$$\theta = \frac{360}{2\pi}$$

On a calculator we find $\theta \approx 57.29577951°$, or $57°\ 17'\ 45''$.

For the more common measures of angles, it is a good idea to memorize the following equivalent degree and radian measures. If you keep in mind that 180° is π radians, the rest of the table will be very easy to remember.

Degrees	Radians
0°	0
30°	$\pi/6$
45°	$\pi/4$
60°	$\pi/3$
90°	$\pi/2$
180°	π
270°	$3\pi/2$
360°	2π

We now relate the radian measure of an angle to a circle to find the arc length. Let s = the length of an arc, and let θ = the angle measured in radians. Then

$$\text{angle in revolutions} = \frac{s}{2\pi r},$$

since one revolution has an arc length (circumference of the circle) of $2\pi r$. Also,

$$\text{angle in radians} = (\text{angle in revolutions})(2\pi).$$

Substituting:

$$\theta = \frac{s}{2\pi r}(2\pi)$$

$$= \frac{s}{r}.$$

This result says that the arc length, s, cut by a central angle θ, *measured in radians,* of a circle of radius r is given by

$$s = r\theta.$$

EXAMPLE: The length of the arc subtended (cut off) by a central angle of $36°$ in a circle with a radius of 20 centimeters (cm) is found by first changing $36°$ to radians so that we can use the formula given above.

$$\frac{36}{360} = \frac{\theta}{2\pi}$$

Solving for θ, we find

$$\frac{\pi}{5} = \theta.$$

Thus,

$$s = 20\left(\frac{\pi}{5}\right)$$

$$= 4\pi.$$

The length of the arc is 4π cm.

PROBLEM SET 1.2

A Problems

1. From memory, give the radian measure for each of the angles whose degree measure is stated.

   a. $30°$ b. $90°$ c. $270°$ d. $45°$ e. $360°$ f. $60°$ g. $180°$

2. From memory, give the degree measure for each of the angles whose radian measure is stated.

   a. π b. $\pi/4$ c. $\pi/3$ d. 2π e. $\pi/2$ f. $3\pi/2$ g. $\pi/6$

3. *Without* first changing to degrees, sketch the angle whose measure is given (remember that π measures a straight angle).

   a. $\pi/2$ b. $2\pi/3$ c. $-\pi/4$ d. 2.5 e. -1

4. *Without* first changing to degrees, sketch the angle whose measure is given (remember that π measures a straight angle).

   a. $-3\pi/2$ b. $\pi/6$ c. $3\pi/4$ d. 6 e. -1.2365

Find the distance between the points given in Problems 5–10.

5. (4, 5) and (8, 8)   6. (6, −2) and (1, 10)   7. (−2, 3) and (5, 10)
8. (−3, −1) and (−6, −4)   9. (4, 3) and (−2, 5)   10. (1, 6) and (5, −8)

Determine whether the points in Problems 11–18 lie on the unit circle.

EXAMPLE: $(\frac{1}{2}, \sqrt{3}/2)$; substitute into the equation of the unit circle:

$x^2 + y^2 = 1;$

$\left(\frac{1}{2}\right)^2 + \left(\frac{\sqrt{3}}{2}\right)^2 \stackrel{?}{=} 1;$

$\frac{1}{4} + \frac{3}{4} \stackrel{?}{=} 1.$

Since the equation is true, $(\frac{1}{2}, \sqrt{3}/2)$ lies on the unit circle.

11. $(\sqrt{2}/2, \sqrt{2}/2)$ 12. $(\sqrt{3}, \sqrt{3}/3)$
13. $(2 + \sqrt{3}, 2 - \sqrt{3})$ 14. (3/5, 4/5)
15. $(\sqrt{3}/2, -1/2)$ 16. $(-\sqrt{3}/2, -1/2)$
17. $\left(\frac{\sqrt{6} + \sqrt{2}}{4}, \frac{\sqrt{6} - \sqrt{2}}{4}\right)$ 18. $\left(\frac{\sqrt{6} - \sqrt{2}}{4}, \frac{-\sqrt{6} - \sqrt{2}}{4}\right)$

Find a positive angle less than one revolution that is coterminal with the angles in Problems 19–32.

19. 400°   20. 540°   21. 750°   22. 1050°   23. 3π   24. 13π/6
25. 11π/3   26. 17π/4   27. −30°   28. −200°   29. −55°   30. −320°
31. −π/4   32. −π

Change the given measures of the angles in Problems 33–40 to decimal degrees to the nearest hundredth.

EXAMPLE: 14° 20′. The problem is to change 20′ to a decimal number. This means that we have 20/60, which by division gives a decimal equivalent of 0.333.... To the nearest hundredth, 14° 20′ = 14.33°.

33. 52° 30′   34. 65° 40′   35. 146° 50′   36. −85° 20′
37. −127° 10′   38. 315° 25′   39. 16° 42′   40. 29° 17′

B Problems

Change the given measures of the angles in Problems 41–45 to decimal degrees to the nearest thousandth.

EXAMPLE: $42° \, 13' \, 40''$. Since $1'' = 1/3600$ of a degree, we have

$$13' \, 40'' = \frac{13}{60} + \frac{40}{3600}$$

$$= \frac{820}{3600}.$$

By division, $13' \, 40'' = 0.22777\ldots°$. To the nearest thousandth, $42° \, 13' \, 40'' = 42.228°$. As you can see, it is quite appropriate to use a calculator to work these problems. If you have a four-function $(+, -, \times, \div)$ calculator, you can do the problem as follows:

For a calculator with algebraic logic, press the following keys:

The desired answer is now displayed.

For a calculator with reverse polish logic, press the following keys:

However, many calculators have a single key conversion for this problem. Check the operating guide for your own calculator.

41. $14° \, 30' \, 50''$ 42. $48° \, 28' \, 10''$ 43. $12' \, 24''$

44. $-94° \, 21' \, 31''$ 45. $281° \, 31' \, 36''$

Find the distance between the points given in Problems 46–51.

46. $(\sqrt{2}, \sqrt{28})$ and $(\sqrt{8}, \sqrt{7})$ 47. $(\sqrt{3}/2, -\frac{1}{2})$ and $(\frac{1}{2}, \sqrt{3}/2)$

48. $(1, 0)$ and $(\sqrt{2}/2, \sqrt{2}/2)$ 49. $(x, f(x))$ and $(2x, 0)$

50. $\left(\dfrac{1}{x-1}, \dfrac{1}{x}\right)$ and $\left(\dfrac{1}{x}, \dfrac{1}{x-1}\right)$; $x > 1$.   51. (α, θ) and (β, ω)

52. Verify that the point

$$\left(\frac{a}{\sqrt{a^2 + b^2}}, \frac{b}{\sqrt{a^2 + b^2}}\right)$$

lies on the unit circle whenever $a^2 + b^2 \neq 0$.

Change the angles in Problems 53–58 to degrees.

EXAMPLE: $\pi/9$. Since

$$\frac{(\text{angle in degrees})}{360} = \frac{(\text{angle in radians})}{2\pi},$$

$$\theta = \frac{360}{2\pi}\text{(angle in radians)}$$

$$= \frac{180}{\pi}\left(\frac{\pi}{9}\right)$$

$$= 20.$$

Thus, $\pi/9$ is an angle with a measure of $20°$.

53. $2\pi/9$ 54. $\pi/10$ 55. $\pi/30$ 56. $5\pi/3$ 57. $-11\pi/12$ 58. $3\pi/18$

Change the angles in Problems 59–64 to degrees correct to the nearest hundredth.

EXAMPLE: 2.3

$$\theta = \frac{180}{\pi}(2.3)$$

$$\approx (57.30)(2.3)$$

$$= 131.79°$$

If you have a calculator, you can obtain a much more accurate answer by using a better approximation for $180/\pi$.

59. 2   60. -3 61. -0.25 62. -2.5 63. 0.4   64. 0.51

Change the angles in Problems 65–70 to radians.

EXAMPLE: 125°

$$\frac{\text{angle in degrees}}{360} = \frac{\text{angle in radians}}{2\pi}$$

$$\theta = \frac{2\pi}{360}\text{(angle in degrees)}$$

$$= \frac{\pi}{180}(125)$$

$$= \frac{25\pi}{36}$$

65. $40°$ 66. $20°$ 67. $-64°$ 68. $-220°$ 69. $254°$ 70. $85°$

Change the angles in Problems 71–74 to radians correct to the nearest hundredth.

EXAMPLE: 43°

$$\theta = \frac{\pi}{180}(43)$$

> $\approx (0.0175)(43)$
>
> $= 0.75$
>
> If you have a calculator, you can obtain a much more accurate answer by using a better approximation for $\pi/180$.

71. 112°   72. 314°   73. −62.8°   74. 350°

In Problems 75–82 find the intercepted arc to the nearest hundredth if the central angle and radius are as given.

75. Angle 1, radius 1 m   76. Angle 2.34, radius 6 cm

77. Angle 3.14, radius 10 m   78. Angle $\pi/3$, radius 4 m

79. Angle $3\pi/2$, radius 15 cm   80. Angle 40°, radius 7 ft

81. Angle 72°, radius 10 ft   82. Angle 112°, radius 7.2 cm

83. How far does the tip of an hour hand measuring 2 cm move in three hours?

84. A 50-cm pendulum on a clock swings through an angle of 100°. How far does the tip travel in one arc?

85. In about 230 B.C. a mathematician named Eratosthenes estimated the radius of the earth using the following information. Syene and Alexandria in Egypt are on the same line of longitude. They are also 500 miles apart. At noon on the longest day of the year, when the sun was directly overhead in Syene, Eratosthenes measured the sun to be 7.2° from the vertical in Alexandria. Because of the distance of the sun, he assumed that the rays were parallel. Thus he concluded that the central angle subtending rays from the center of the earth to Syene and Alexandria was also 7.2°. Using this information, find the approximate radius of the earth.

86. *Calculator Problem.* Omaha, Nebraska, is located at approximately 97° W longitude, 41° N latitude; Wichita, Kansas, is located at approximately 97° W longitude, 37° N latitude. Notice that these two cities have about the same longitude. If we know that the radius of the earth is about 6370 kilometers (km), what is the distance between these cities to the nearst 10 km?

87. *Calculator Problem.* Entebbe, Uganda, is located at approximately 33°E longitude, and Stanley Falls in Zaire is located at 25°E longitude. Also, both these cities lie approximately on the equator. If we know that the radius of the earth is about 6370 km, what is the distance between the cities to the nearest 10 km?

Mind Bogglers

88. *Calculator Problem.* Suppose it is known that the moon subtends an angle of 45.75′ at the center of the earth. It is also known that the center of the moon is 238,866 miles from the surface of the earth. What is the diameter of the moon to the nearest 10 miles?

89. One side of a triangle is 20 cm longer than another, and the angle between them is 60°. If two circles are drawn with these sides as diameters, one of the points of intersection of the two circles is the common vertex. How far from the third side is the other point of intersection?

1.3 GRAPHING LINEAR AND QUADRATIC FUNCTIONS (*Get the Picture*)

We now review two functions from algebra: linear and quadratic functions.

Definition of Linear and Quadratic Functions: A function f is called

1. **linear** if $f(x) = mx + b$;
2. **quadratic** if $f(x) = ax^2 + bx + c$, $a \neq 0$.

In Section 1.1 we spoke of the graph of a relation. This is an essential notion in mathematics—one that relates the set of ordered pairs of a relation to the set of ordered pairs of a curve in a one-to-one fashion. Thus simple curves can usually be determined by plotting points. But plotting points at random is generally not efficient.

We will consider linear and quadratic functions from a viewpoint different from the one you used in algebra. Hopefully these methods of graphing will generalize more easily to the graphing of new functions you will meet in this course.

LINES

In algebra you studied several forms of the equation of a line.

Standard form: $Ax + By + C = 0$, where (x, y) is any point on the line and A, B, and C are constants, A and B not both 0.

Slope-intercept form: $y = mx + b$, where m is the slope and b is the y-intercept.

Point-slope form: $y - y_1 = m(x - x_1)$, where (x_1, y_1) is a point on the line.

Two-point form: $y - y_1 = \left(\dfrac{y_2 - y_1}{x_2 - x_1} \right)(x - x_1)$, where (x_1, y_1) and (x_2, y_2) are points on the line.

In this section we will focus attention on linear equations of the form $y = mx$, and in the next section we'll consider variations of this form. First, notice that $y = mx$ contains the origin, since $(0, 0)$ satisfies the equation. Next, we determine the line by considering the *slope, m*. Recall that the slope of a line represents its "steepness" and is often defined by selecting two points on the line—say (x_1, y_1) and (x_2, y_2), where $x_1 \neq x_2$—as shown in Figure 1.7. Then:

$$m = \frac{\text{vertical change}}{\text{horizontal change}} = \frac{y_2 - y_1}{x_2 - x_1} = \frac{\text{rise}}{\text{run}}.$$

Since the slope of a line is a constant, the ratio of rise to run will be the same regardless of the two distinct points chosen, as can be seen by recalling a result

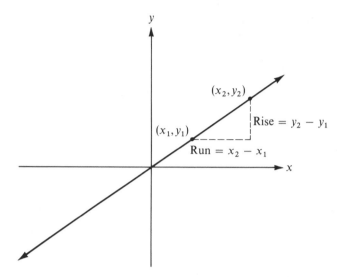

FIGURE 1.7   Slope of a line

from geometry. Consider a line passing through $A(x_1, y_1)$, $B(x_2, y_2)$, $A'(x_3, y_3)$, and $B'(x_4, y_4)$. Also consider $C(x_2, y_1)$ and $C'(x_4, y_3)$, as shown in Figure 1.8. Since $\angle C$ and $\angle C'$ are right angles (why?), we see that $\angle C \cong \angle C'$ (recall that $\cong$ means "is congruent to"). Also, $\overline{AC}$ and $\overline{A'C'}$ are parallel to the x-axis (why?), so they are parallel lines with $\overleftrightarrow{AB}$ a transversal. This means that $\angle BAC$ and $\angle B'A'C'$ are corresponding angles such that $\angle BAC \cong \angle B'A'C'$. If two angles

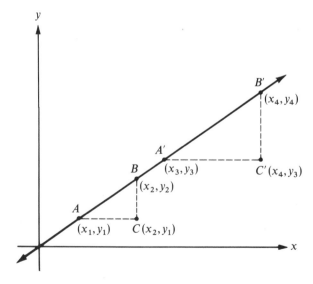

FIGURE 1.8   Four arbitrary points A, B, A', and B' on a given line

of one triangle are congruent to two angles of another triangle, the triangles are similar.

$$\triangle ABC \sim \triangle A'B'C'$$

(The symbol $\sim$ means "is similar to"). Since corresponding parts of similar triangles are proportional,

$$\frac{BC}{AC} = \frac{B'C'}{A'C'}$$

or, in terms of the coordinates,

$$\frac{y_2 - y_1}{x_2 - x_1} = \frac{y_4 - y_3}{x_4 - x_3}.$$

Each of these ratios represents the slope; since they are equal, the slope of a line is the same regardless of the points chosen. In practice, then, we can select those points on the line that are most convenient for the problem at hand.

EXAMPLES:

1. Graph $y = (-5/3)x$. This line passes through the origin and has slope $m = -5/3$. To graph, we start at the known point (the origin) and count out the slope using rise $= -5$ and run $= 3$, as shown in Figure 1.9.

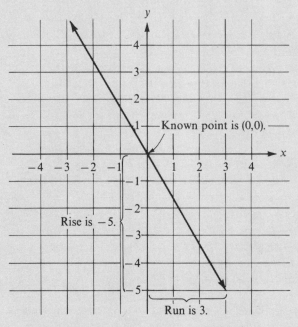

FIGURE 1.9   Negative slope: $y = (-5/3)x$

2. Graph $6x - 2y = 0$. Solve for y:

$$y = 3x.$$

This line passes through the origin and has slope $m = 3$. Since $3 = 3/1$, we use rise = 3 and run = 1, as shown in Figure 1.10.

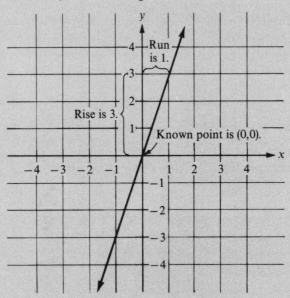

FIGURE 1.10   Positive slope: $y = 3x$

3. Graph $y = \frac{2}{3}x$ on the coordinate axes whose scales are such that each square represents:

a. 1 unit.   b. $\frac{1}{5}$ unit.   c. 500 units.

The solutions are shown in Figure 1.11.

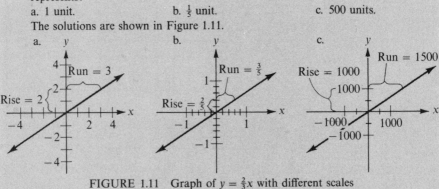

FIGURE 1.11   Graph of $y = \frac{2}{3}x$ with different scales

For Example 3 above, notice that $m = \frac{2}{3}$. For part (a) we used rise = 2, run = 3; for part (b), rise = $\frac{2}{5}$, run = $\frac{3}{5}$; for part (c), rise = 1000, run = 1500. For each part the *ratio*

$$\frac{\text{rise}}{\text{run}} = \frac{2}{3} = \frac{2/5}{3/5} = \frac{1000}{1500}.$$

$$\phantom{\frac{\text{rise}}{\text{run}} = }\text{(a)}\quad\text{(b)}\quad\text{(c)}$$

Thus, if you are given a ratio

$$\frac{y}{x} = \frac{2}{3},$$

you *cannot* say $y = 2$ and $x = 3$, since there are *many* choices of x and y that satisfy this ratio. (See Figure 1.11; the *line* includes all possible x and y values satisfying $y/x = \frac{2}{3}$.)

PARABOLAS

In algebra you found that, if $a \neq 0$, the graph of $y = ax^2$ is a parabola. As with the line, we first notice that this parabola passes through the origin, since $(0, 0)$ satisfies the equation (regardless of the value for a):

$$0 = a(0).$$

Next, we determine the parabola by considering different choices for a. This number a indicates the steepness of the parabola, which we see from the following examples.

EXAMPLES:

1. Graph the following on the same coordinate axes: (a) $y = x^2$; (b) $y = 2x^2$; (c) $y = \frac{1}{4}x^2$. We do so by considering a table of values and plotting the points, as in Figure 1.12. After we plot the points in column (a), we connect them to form a curve. Do the same for columns (b) and (c).

	x	(a) $y = x^2$	(b) $y = 2x^2$	(c) $y = \frac{1}{4}x^2$
positive x-values	1	1	2	$\frac{1}{4}$
	2	4	8	1
	3	9	18	$\frac{9}{4}$
	4	16	32	4
negative x-values	-1	1	2	$\frac{1}{4}$
	-2	4	8	1
	-3	9	18	$\frac{9}{4}$
	-4	16	32	4

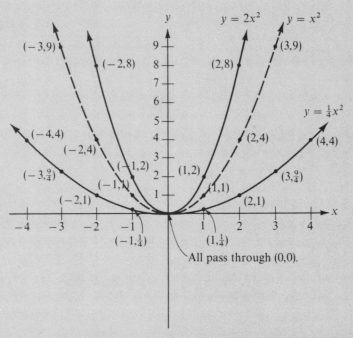

FIGURE 1.12

2. Graph the following on the same coordinate axes: (a) $y = -x^2$; (b) $y = -2x^2$; (c) $y = -\frac{1}{4}x^2$. We graph them by considering a table of values. These points are plotted in Figure 1.13.

	x	(a) $y = -x^2$	(b) $y = -2x^2$	(c) $y = -\frac{1}{4}x^2$
positive x-values	1	-1	-2	$-\frac{1}{4}$
	2	-4	-8	-1
	3	-9	-18	$-\frac{9}{4}$
	4	-16	-32	-4
negative x-values	-1	-1	-2	$-\frac{1}{4}$
	-2	-4	-8	-1
	-3	-9	-18	$-\frac{9}{4}$
	-4	-16	-32	-4

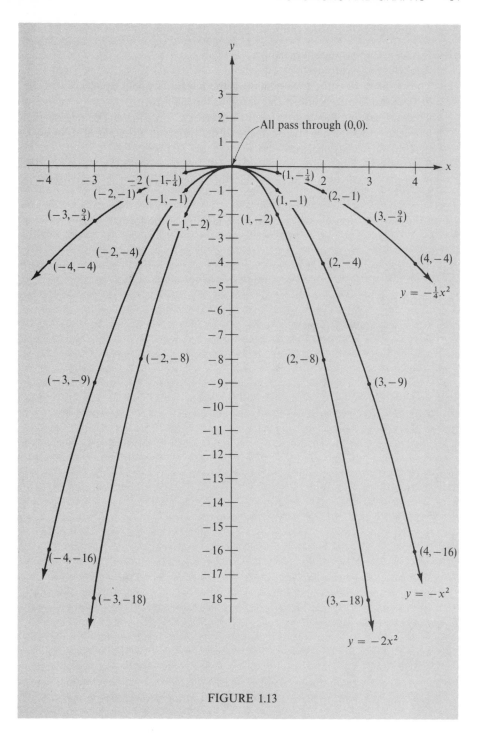

FIGURE 1.13

We can make four observations based on the examples.

1. If $a > 0$, the parabola opens up;
   if $a < 0$, the parabola opens down.
2. The point $(0, 0)$ is the lowest point if the parabola opens up;
   $(0, 0)$ is the highest point if the parabola opens down.
3. The vertical line passing through the vertex is the y-axis. The points to the left of this line form a mirror image of the points to the right of it, so we say that the curve is *symmetric* with respect to this line.
4. Relative to a fixed scale, the magnitude of a determines the "fatness" of the parabola; small values of $|a|$ yield "fat" parabolas, and large values of $|a|$ yield "skinny" parabolas.

In the next section we will consider variations of the form $y = ax^2$.

PROBLEM SET 1.3

A Problems

Graph the equations given in Problems 1–18.

1. $y = \frac{2}{3}x$
2. $y = \frac{3}{4}x$
3. $y = -\frac{1}{2}x$
4. $y = -\frac{4}{5}x$
5. $y = 2x$
6. $y = 3x$
7. $y = x$
8. $y = -x$
9. $y = 3x^2$
10. $y = \frac{1}{10}x^2$
11. $y = \frac{1}{100}x^2$
12. $y = 5x^2$
13. $y = -3x^2$
14. $y = -\frac{1}{10}x^2$
15. $y = -\frac{1}{100}x^2$
16. $y = -5x^2$
17. $y = \frac{2}{3}x^2$
18. $y = -\frac{4}{5}x^2$

B Problems

Graph the equations given in Problems 19–24.

19. $2x + 3y = 0$
20. $2x^2 + 3y = 0$
21. $\pi x - y = 0$
22. $\pi x^2 - y = 0$
23. $\sqrt{5}x + 3y = 0$
24. $\sqrt{5}x^2 - 3y = 0$

Graph the equations given in Problems 25–36 with the given restrictions.

EXAMPLE: $5x - 4y = 0$: $-2 \leq x < 3$

Step 1: Graph the given curve, $y = (5/4)x$. The result is shown by a dotted line in Figure 1.14.

Step 2: The restrictions might be in terms of x or in terms of y. In this example they are in terms of x. We locate these values as shown by the arrows in Figure 1.14.

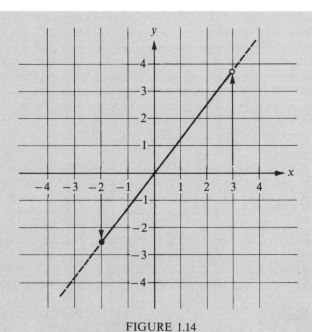

FIGURE 1.14

Step 3: We draw that part of the curve that lies between these x-values. The result is shown by a solid line in Figure 1.14. Notice that $x = -2$ is included, indicated by ●, and $x = 3$ is excluded, indicated by ○.

25. $x - y = 0$: $-3 \le x \le 4$ 26. $2x + 5y = 0$: $-4 < x \le 3$

27. $5x - y = 0$: $1 < y < 4$ 28. $x + 4y = 0$: $-3 < y < 4$

29. $x^2 - y = 0$: $-3 < x < 3$ 30. $2x^2 + y = 0$: $-4 \le x \le 2$

31. $x^2 + \frac{1}{2}y = 0$: $y \le -1$ 32. $x^2 - 3y = 0$: $y \le 2$

33. $5x + 2y = 0$: $-0.1 \le x \le 0.1$ 34. $5x + 2y = 0$: $100 \le x \le 500$

35. $y = -5x$: $-\frac{1}{5} \le x \le \frac{1}{5}$ 36. $y = x^2$: $-\frac{1}{10} \le x \le \frac{1}{10}$

37. Graph $y = x$. Let $P_1(2, 2)$, $Q_1(2, 0)$, $P_2(5, 5)$, $Q_2(5, 0)$, and $O(0, 0)$. Show that $\triangle P_1OQ_1 \sim \triangle P_2OQ_2$.

38. Graph $y = \frac{2}{5}x$. Let $P_1(5, 2)$, $Q_1(5, 0)$, $P_2(10, 4)$, $Q_2(10, 0)$, and $O(0, 0)$. Show that $\triangle P_1OQ_1 \sim \triangle P_2OQ_2$.

39. Graph $y = -\frac{2}{3}x$. Choose *any* two points P_1 and P_2 on the line, and draw perpendiculars from these points to the x-axis at Q_1 and Q_2, respectively. If O is the origin, show that $\triangle P_1OQ_1 \sim \triangle P_2OQ_2$.

40. Let $y = mx$ be any line passing through $O(0, 0)$, and let $P_1(x_1, y_1)$ and $P_2(x_2, y_2)$ be any points on the line. Draw segments from P_1 and P_2 perpendicular to the x-axis. Show that the resulting triangles are similar.

41. (*Continuation of Problem 40.*) Let $|OP_1| = r_1$ and $|OP_2| = r_2$. Show that:

$$\frac{x_1}{r_1} = \frac{x_2}{r_2}, \qquad \frac{y_1}{r_1} = \frac{y_2}{r_2}, \qquad \frac{y_1}{x_1} = \frac{y_2}{x_2}.$$

Mind Boggler

42. An investment syndicate owns a parcel of land that is 1 km square. They decide to subdivide the parcel into five parts, each with equal area, but for aesthetic reasons do not want to divide the land into parallel strips. Can this be done?

1.4 TRANSLATIONS WITH FUNCTIONS (*Can Mathematics Be Translated?*)

In the last section we considered functions defined by

$$f(x) = mx;$$
$$f(x) = ax^2, \ a \neq 0.$$

In this section we'll consider translations of these functions. By a *translation* we mean a shift of the entire curve to the left, right, up, or down.

An up or down translation of k units has the form $y - k = f(x)$, as shown in Figure 1.15. If k is positive, the shift is up; if k is negative, the shift is down. However, since the form is $y - k$, you must be careful to interpret k correctly.

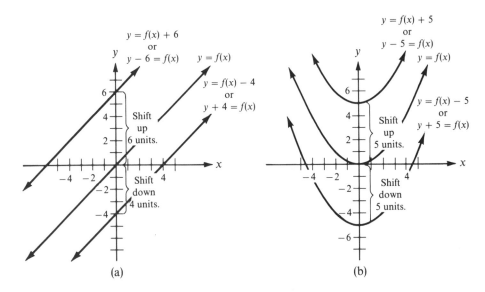

FIGURE 1.15   Up or down translation of the form $y - k = f(x)$

EXAMPLES: What is k in the following equations?

1. $y - 3 = f(x)$; $k = 3$

2. $y - 5 = f(x)$; $k = 5$

3. $y + 2 = f(x)$;   since $y + 2 = y - (-2)$, we see that $k = -2$.

4. $y + 5 = f(x)$; $k = -5$

5. $y = f(x)$; $k = 0$

6. $y - 2 = f(x)$; $k = 2$

7. $y = f(x) - 4$;   writing $y + 4 = f(x)$, we see that $k = -4$.

A right or left translation of h units has the form $y = f(x - h)$, as shown in Figure 1.16. If h is positive, the shift is to the right; if h is negative, the shift is to the left.

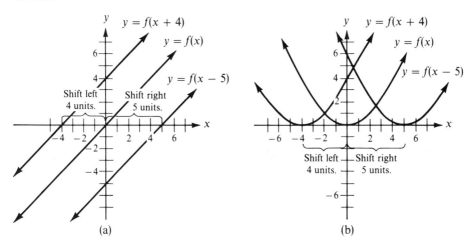

FIGURE 1.16   Right or left translation of the form $y = f(x - h)$

In general, if $y = f(x)$ is some curve, then

$$y - k = f(x - h)$$

is a translation that shifts the curve h units to the right (h positive) or left (h negative) and k units up (k positive) or down (k negative).

EXAMPLES: Find (h, k) for the following equations.

1. $y - 5 = f(x - 7)$; $(h, k) = (7, 5)$

2. $y + 6 = f(x - 1)$; $(h, k) = (1, -6)$

3. $y + 1 = f(x + 3)$; $(h, k) = (-3, -1)$

4. $y = f(x)$; $(h, k) = (0, 0)$. This indicates no shift.

5. $y - 15 = f(x + \frac{1}{3})$; $(h, k) = (-\frac{1}{3}, 15)$

6. $y - 6 = f(x) + 15$; write the equation as $y - 21 = f(x)$; thus, $(h, k) = (0, 21)$. This indicates no horizontal shift.

To sketch a curve of the form $y - k = f(x - h)$, we can first sketch $y = f(x)$ and then locate the point (h, k). From this point we can trace out the identical curve on which each point has been translated the same distance.

EXAMPLES: Compare each equation to the form $y = mx$ and graph each by making an appropriate shift.

1. $y - 3 = \frac{1}{2}x$: By inspection, this is a line with the form $y = \frac{1}{2}x$, where $(h, k) = (0, 3)$. The shift is up three units. The sketch is shown in Figure 1.17.

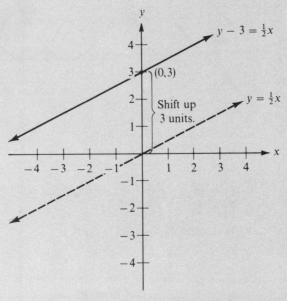

FIGURE 1.17   Graph of $y - 3 = \frac{1}{2}x$

2. $y = 2(x - 3)$: By inspection, this is a line with the form $y = 2x$, where $(h, k) = (3, 0)$. The shift is three units to the right, as shown in Figure 1.18.

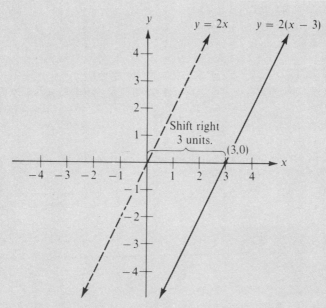

FIGURE 1.18   Graph of $y = 2(x - 3)$

3. $y - 5 = \frac{2}{3}(x - 2)$: By inspection, this is a line with the form $y = \frac{2}{3}x$, where (h, k) = (2, 5). This means that (2, 5) now plays the role of the origin. The graph is shown in Figure 1.19.

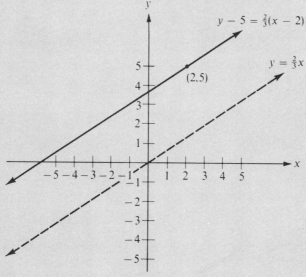

FIGURE 1.19   Graph of $y - 5 = \frac{2}{3}(x - 2)$

4. $y + 3 = -\frac{2}{3}(x + 7)$: *Mentally* draw the line $y = -\frac{2}{3}x$. Since $(h, k) = (-7, -3)$, we draw the line with slope $-\frac{2}{3}$ passing through $(-7, -3)$, as shown in Figure 1.20.

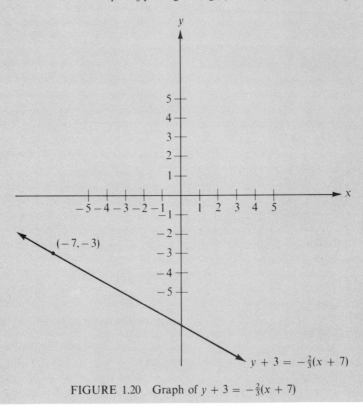

FIGURE 1.20   Graph of $y + 3 = -\frac{2}{3}(x + 7)$

Actually, what we've done here is basically what you did in algebra. We've simply introduced the idea of shifting the curve from a standard position. You may ask why we would want to do this, since we can draw the line without ever mentioning shifts. The reason is that this idea of shifting a curve will generalize for any function and can be used in further study in this and other courses.

EXAMPLES:

1. $y - 3 = (x - 4)^2$: This equation has the form $y - k = (x - h)^2$, which is a parabola for which $(h, k) = (4, 3)$. The standard-position parabola is $y = x^2$. Next, plot $(4, 3)$, which defines the shift. The point (h, k) is the *vertex* of the parabola. Next, draw the parabola $y = x^2$ at the vertex $(4, 3)$. The result is shown in Figure 1.21.

2. $y + 2 = \frac{1}{2}(x + 5)^2$: By inspection, $(h, k) = (-5, -2)$, and the standard-position parabola is $y = \frac{1}{2}x^2$. The result is shown in Figure 1.22.

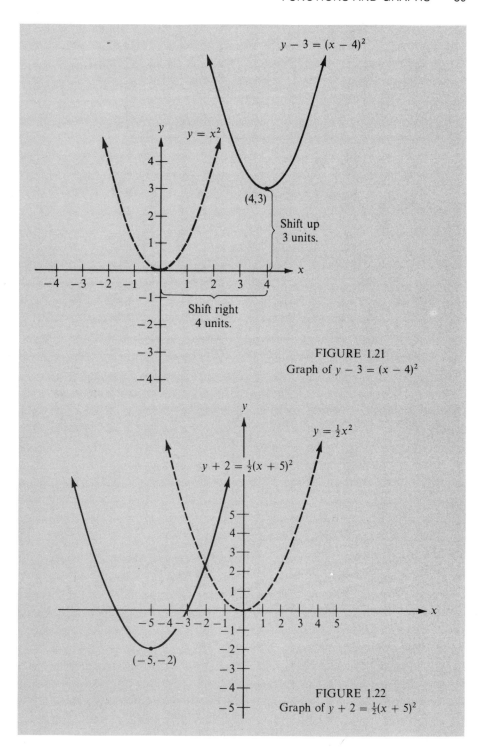

$y - 3 = (x - 4)^2$

$y = x^2$

(4,3)

Shift up
3 units.

Shift right
4 units.

FIGURE 1.21
Graph of $y - 3 = (x - 4)^2$

$y = \frac{1}{2}x^2$

$y + 2 = \frac{1}{2}(x + 5)^2$

$(-5, -2)$

FIGURE 1.22
Graph of $y + 2 = \frac{1}{2}(x + 5)^2$

Summary: Given $y = f(x)$:

    right shift h units (*h* positive): $y = f(x - h)$

    left shift h units (*h* negative): $y = f(x - h)$

    up shift k units (*k* positive): $y - k = f(x)$

    down shift k units (*k* negative): $y - k = f(x)$

In general, $y - k = f(x - h)$ passes through (h, k) and is the graph of $y = f(x)$, which has been shifted *h* units horizontally and *k* units vertically.

PROBLEM SET 1.4

A Problems

Graph each pair of curves in Problems 1–20 on the same axes.

1. $y = x$; $y - 2 = (x - 5)$
2. $y = \frac{2}{3}x$; $y - 2 = \frac{2}{3}(x - 5)$
3. $y = -\frac{1}{2}x$; $y - 3 = -\frac{1}{2}(x + 2)$
4. $y = -\frac{2}{3}x$; $y - 4 = -\frac{2}{3}(x + 5)$
5. $y = \frac{3}{4}x$; $y + 1 = \frac{3}{4}(x + 6)$
6. $y = \frac{4}{5}x$; $y + 5 = \frac{4}{5}(x - 7)$
7. $y = x^2$; $y + 3 = x^2$
8. $y = x^2$; $y = (x - 6)^2$
9. $y = x^2$; $y = (x - 4)^2$
10. $y = x^2$; $y = (x + 4)^2$
11. $y = \frac{2}{3}x^2$; $y - 1 = \frac{2}{3}(x - 5)^2$
12. $y = \frac{4}{5}x^2$; $y - 1 = \frac{4}{5}(x + 3)^2$
13. $y = 3x$; $y - \sqrt{2} = 3(x + \sqrt{5})$
14. $y = 3x^2$; $y - \sqrt{2} = 3(x + \sqrt{5})^2$
15. $y = \frac{2}{5}x$; $y - \pi = \frac{2}{5}x$
16. $y = -\frac{2}{5}x$; $y = -\frac{2}{5}(x + \pi)$
17. $y = \frac{2}{5}x^2$; $y - \pi = \frac{2}{5}x^2$
18. $y = -\frac{2}{5}x^2$; $y = -\frac{2}{5}(x - \pi)^2$
19. $y = 0.1x$; $y - \pi = 0.1(x + \pi/2)$
20. $y = 0.1x^2$; $y - \pi = 0.1(x + \pi/6)^2$

Let f be the function graphed in Figure 1.23. Graph Problems 21–32.

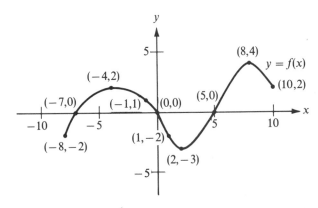

FIGURE 1.23   Graph of a function $y = f(x)$

21. $y + 3 = f(x)$ 22. $y - 5 = f(x)$ 23. $y = f(x - 4)$

24. $y = f(x + 2)$ 25. $y - 2 = f(x - 1)$ 26. $y + 4 = f(x - 3)$

B Problems

27. $y = f(x + \pi)$ 28. $y + \pi = f(x)$ 29. $y = \frac{1}{2}f(x)$

30. $y = -f(x)$ 31. $y = 2f(x)$ 32. $y = -2f(x)$

33. An *even* function f is one for which $f(-x) = f(x)$. For example, if $f(x) = 2x^2$, then $f(-x) = 2(-x)^2 = 2x^2 = f(x)$; thus, $f(x) = 2x^2$ is even. On the other hand, if $g(x) = 2x^3$, then $g(-x) = 2(-x)^3 = 2(-x^3) = -2x^3$; thus, g is not an even function. Which of the following define even functions?

a. $f_1(x) = x^2 + 1$ b. $f_2(x) = |x|$ c. $f_3(x) = \dfrac{1}{3x^3 - 4}$

d. $f_4(x) = x^3 + x$ e. $f_5(x) = \dfrac{1}{|x^3 + 3|}$ f. $f_6(x) = \dfrac{1}{|x^3 + x|}$

34. An *odd* function f is one for which $f(-x) = -f(x)$. For example, for $f(x) = 2x^2$ and $g(x) = 2x^3$ of Problem 33, we see that g is an odd function. Which of the functions defined in Problem 33 are odd functions?

35. a. Graph $y = (x + 1)$ and $y = -(x + 1)$ on the same axes.
 b. Graph $y = (x + 1)^2$ and $y = -(x + 1)^2$ on the same axes.
 c. Make a conjecture about the relationship between $y = f(x)$ and $y = -f(x)$.
 d. Test your conjecture by graphing $y - 3 = \frac{2}{3}(x + 1)^2$ and $y - 3 = -\frac{2}{3}(x + 1)^2$.

36. Let $f_1(x) = x^2$ and $f_2(x) = 6x + 9$.
 a. Graph $f_1(x)$ and $f_2(x)$ on the same axes.
 b. Graph $g(x) = f_1(x) + f_2(x) = x^2 + 6x + 9$
$$= (x + 3)^2$$
 on the same axes you used for part (a).
 c. Find the following values of $f_1(x)$, $f_2(x)$, and $g(x)$:
  i. $x = 0$ ii. $x = 1$ iii. $x = -1$ iv. $x = 2$ v. $x = -2$
  Plot these points on the same axes you used for part (a).
 d. Make a conjecture about how $g(x)$ could have been graphed by adding the functions $f_1(x)$ and $f_2(x)$.

37. Graph $g(x) = x^2 - 4x + 4$ by graphing $f_1(x) + f_2(x)$, where $f_1(x) = x^2$ and $f_2(x) = -4x + 4$ (see Problem 36). Check your answer by graphing $g(x) = (x - 2)^2$.

38. Graph $g(x) = \frac{2}{3}x^2 + 5x - 1$ by graphing $f_1(x) = \frac{2}{3}x^2$ and $f_2(x) = 5x - 1$ (see Problem 36).

39. a. Graph $f(x) = 2x + 3$.
 b. Graph $g(x) = (x - 3)/2$.
 c. $f[g(x)]$ means $f[(x - 3)/2]$, since $g(x) = (x - 3)/2$. Find $f[g(x)]$.
 d. $g[f(x)]$ means $g(2x + 3)$, since $f(x) = 2x + 3$. Find $g[f(x)]$.
 e. Does $f[g(x)]$ always equal $g[f(x)]$ for *any* functions f and g?
 f. Find some other examples for which

$$f[g(x)] = g[f(x)],$$

 and then make a conjecture about the nature of the functions f and g whenever this relationship holds.

Set up the coordinate axes for Problems 40–49 so that $-15 \le x \le 15$ and $-5 \le y \le 20$, and graph all the problems on the same axes.*

40. $y + 3 = \frac{1}{3}(x + 8)^2$: $-14 \le x \le -8$ 41. $y = 10(x + 8.2)$: $-8 < x < -7$

42. $y - 16 = -\frac{1}{4}(x + 3)^2$: $-7 \le x \le -1$ 43. $y - 14 = -x$: $-1 \le x \le 2$

44. $y - 17 = -\frac{5}{2}x$: $2 < x < 4$ 45. $y - 12 = -\frac{4}{3}(x + \frac{25}{2})^2$: $y > 9$

46. $y + 3 = \frac{3}{16}(x + 3)^2$: $y \le 9$ 47. $y - 12 = -\frac{1}{3}(x - 8)^2$: $y > 9$

48. $y + \frac{13}{3} = \frac{4}{3}(x - 8)^2$: $y \le 4x - 35$

49. Graph the simultaneous solution of:

$$\begin{cases} y - 1 \le \dfrac{-1}{64}(x + 3)^2 \\[2mm] y - \dfrac{1}{4} \ge \dfrac{-1}{64}(x + 3)^2 \\[2mm] y + 3 \ge \dfrac{3}{16}(x + 3)^2 \end{cases}$$

Mind Boggler

50. In Problems 40–49 we drew a picture by writing the equations of lines and parabolas. Now draw another picture using only lines, parabolas, or circles, and then describe your picture with a series of equations or inequalities.

1.5 INVERSE OF A FUNCTION (*Here to There and Back Again*)

Consider the equation $3x = 12$. To solve this equation, we multiply both sides by $\frac{1}{3}$.

$$\left(\frac{1}{3}\right)(3x) = \left(\frac{1}{3}\right)(12)$$
$$1 \cdot x = 4$$
$$x = 4$$

Why do we multiply by $\frac{1}{3}$? The answer, of course, is that $\frac{1}{3}$ is the *inverse* of 3 for multiplication. Also, recall that $\frac{1}{3}$ can be written as 3^{-1}. Thus, if we start with a number x multiplied by 3, and then wish to "undo" the operation of multiplying by 3, we simply multiply by the inverse of 3—namely $\frac{1}{3}$.

This simple example is introduced here because we wish to generalize the idea to functions. That is, given a number x, we can evaluate some function f at x. Next, we "undo" the procedure to obtain x again (that is, to get back to

*My thanks to Pat Boyle of Santa Rosa Junior College for these problems.

where we started). Let's consider an example. Let $f(x) = 2x + 3$, where $D = \{0, 1, 3, 5\}$. Then $f = \{(0, 3), (1, 5), (3, 9), (5, 13)\}$. We can also state the relationships using the following "map."

Domain of f		Range of f
0	$\xrightarrow{\ f\ }$	3
1	$\longrightarrow$	5
3	$\longrightarrow$	9
5	$\longrightarrow$	13

To "undo" the results of f, we define a new relation so that each element in the range maps back into the original element of the domain. Let's denote this relation by f^{-1}. (Be careful about this notation: f^{-1} means the *inverse* of the function f; it does not mean $1/f$ (1 divided by f).

Domain of f		Range of f Domain of f^{-1}		Range of f^{-1}
0	$\xrightarrow{\ f\ }$	3	$\xrightarrow{\ f^{-1}\ }$	0
1	$\longrightarrow$	5	$\longrightarrow$	1
3	$\longrightarrow$	9	$\longrightarrow$	3
5	$\longrightarrow$	13	$\longrightarrow$	5

The relation f^{-1} can also be written

$$f^{-1} = \{(3, 0), (5, 1), (9, 3), (13, 5)\}.$$

Notice that the inverse of f, which is a set of ordered pairs (x, y), is the set of ordered pairs with the components interchanged—namely (y, x). In terms of the original equation, if

$$f = \{(x, y)\,|\,y = 2x + 3\},$$

then

$$f^{-1} = \{(y, x)\,|\,y = 2x + 3\}.$$

However, because we are unaccustomed to ordered pairs of the form (y, x), we interchange the x and y values to write

$$f^{-1} = \{(x, y)\,|\,x = 2y + 3\}.$$

Solving this latter equation for y, we get:

$$x = 2y + 3;$$
$$x - 3 = 2y;$$
$$y = \frac{x - 3}{2}.$$

If $g(x) = (x - 3)/2$, then g is called the *inverse function* for f. For example, suppose we start with the number 3 and evaluate f:

$$f(3) = 2(3) + 3$$
$$= 9.$$

Next, we apply g to this result:

$$g(9) = \frac{9 - 3}{2}$$
$$= 3.$$

This process can also be written $g[f(3)]$, which is called the *composition of functions*—first f, then g. More generally:

$$g[f(x)] = g[2x + 3]$$
$$= \frac{(2x + 3) - 3}{2}$$
$$= \frac{2x}{2}$$
$$= x.$$

EXAMPLES: Find the inverse of each of the following functions.

1. $f_1 = \{(4, 7), (9, 2), (7, 3), (1, 6)\}$. Then the inverse of f_1 is $\{(7, 4), (2, 9), (3, 7),(6, 1)\}$.

2. $f_2 = \{(6, 1), (7, 2), (8, 1), (9, 3)\}$. Then the inverse of f_2 is $\{(1, 6), (2, 7), (1, 8), (3, 9)\}$.

3. $f_3 = \{(x, y) \mid y = 5x + 4\}$. Then the inverse of f_3 is $\{(y, x) \mid y = 5x + 4\}$. But we would interchange the x and y values to write $\{(x, y) \mid x = 5y + 4\}$.

4. $f_4 = \{(x, y) \mid y = x^2\}$. Then the inverse of f_4 is $\{(x, y) \mid x = y^2\}$.

As you can see from the preceding examples, the inverse of a function is not necessarily a function. The function f_2 has an inverse that is not a function, since the first component, 1, is associated with both 6 and 8. For Example 3, $f_3(x) = 5x + 4$ means that

$$y = 5x + 4.$$

The inverse is

$$x = 5y + 4.$$

To see if this is a function, solve for y:

$$y = \frac{x - 4}{5}.$$

We now see that each x-value yields a single y-value, so the inverse of f_3 is a function. When the inverse is a function, we write $f_3^{-1}(x) = (x - 4)/5$.

If f and g are two given functions, we can check to see if they are inverses by checking $f[g(x)]$ and $g[f(x)]$. If the result of these operations gives x, then f and g are inverses. For Example 3:

$$f_3^{-1}[f_3(x)] = f_3^{-1}[5x + 4] \quad \text{and} \quad f_3[f_3^{-1}(x)] = f_3\left[\frac{x - 4}{5}\right]$$

$$= \frac{(5x + 4) - 4}{5} \qquad\qquad = 5\left(\frac{x - 4}{5}\right) + 4$$

$$= \frac{5x}{5} \qquad\qquad\qquad = (x - 4) + 4$$

$$= x \qquad\qquad\qquad\qquad = x$$

The graphs of f_3 and f_3^{-1} are shown in Figure 1.24.

For Example 4, $f_4(x) = x^2$ means that

$$y = x^2.$$

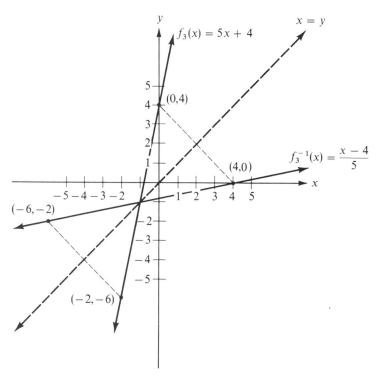

FIGURE 1.24   Graphs of $f_3(x) = 5x + 4$ and $f_3^{-1}(x) = (x - 4)/5$

The inverse is given by

$$x = y^2.$$

Solving for y:

$$y = \pm\sqrt{x},$$

which is not a function, because each positive value of x yields two values of y. Thus, since the inverse of f_4 is not a function, we do not write f_4^{-1}. The graphs of f_4 and its inverse are shown in Figure 1.25.

When working with functions and their inverses, we often find it desirable for the inverse also to be a function. If we restrict the domain of a function, we frequently can force the inverse to be a function too. For example, if

$$F_4(x) = x^2, \quad x \geq 0,$$

then

$$F_4^{-1}(x) = \sqrt{x}$$

is the inverse function. The graphs of F_4 and F_4^{-1} are shown in Figure 1.26. Notice that $f_4 \neq F_4$, since their domains are different. We will agree to use the following convention about notation: if f denotes a function whose inverse is not

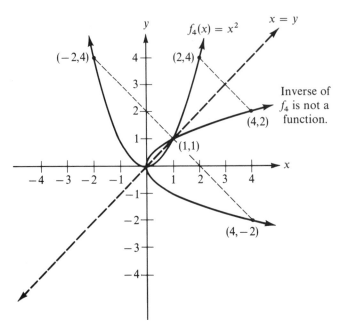

FIGURE 1.25   Graphs of $y = x^2$ and $x = y^2$

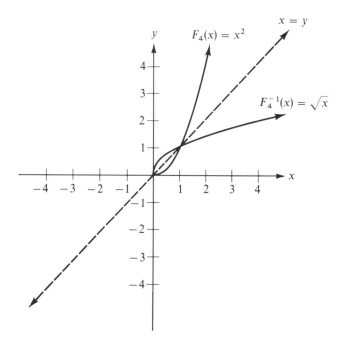

FIGURE 1.26   Graphs of $F_4(x) = x^2$ and $F_4^{-1}(x) = \sqrt{x}$

a function, then F denotes the function f with restrictions on its domain so that its inverse is a function. Consider the function from Example 2:

function	domain	inverse
$f_2 = \{(6,1), (7,2), (8,1), (9,3)\}$	$\{6,7,8,9\}$	$\{(1,6), (2,7), (1,8), (3,9)\}$ This is not a function.
$F_2 = \{(7,2), (8,1), (9,3)\}$	$\{7,8,9\}$	$\{(2,7), (1,8), (3,9)\}$ This is a function.

Notice that, although $f_2 \neq F_2$, they are exactly the same except that the domain of f_2 has been limited so as to force F_2 to have an inverse that is a function.

EXAMPLES: Given a function f, define a function F that meets the following conditions.
i.   If the inverse of f is a function, let $F = f$.
ii. If the inverse of f is not a function, restrict the domain of f in order to define a new function F so that the inverse of F is a function.

1. $f(x) = 2x + 1$, where $D = \{-2, -1, 0, 1, 2\}$.
    $f = \{(-2, -3), (-1, -1), (0, 1), (1, 3), (2, 5)\}$
    $f^{-1} = \{(-3, -2), (-1, -1), (1, 0), (3, 1), (5, 2)\}$, which is a function.
    Define $F = f$.

2. $f(x) = 2x^2 + 1$, where $D = \{-2, -1, 0, 1, 2\}$.
    $f = \{(-2, 9), (-1, 3), (0, 1), (1, 3), (2, 9)\}$
    $f^{-1} = \{(9, -2), (3, -1), (1, 0), (3, 1), (9, 2)\}$, which is not a function.
    Thus we restrict the domain of f (which can usually be done in more than one way): Let $F = f$ where the domain of F is $\{0, 1, 2\}$.
    $F = \{(0, 1), (1, 3), (2, 9)\}$
    $F^{-1} = \{(1, 0), (3, 1), (9, 2)\}$, which is a function.

3. $f(x) = 2x + 1$
    The set of reals is understood to be the domain. Write f as a set of ordered pairs:

    $$f = \{(x, y) \mid y = 2x + 1\}$$
    $$f^{-1} = \{(y, x) \mid y = 2x + 1\} \text{ or } \{(x, y) \mid x = 2y + 1\}$$

    Solving for y:

    $$x = 2y + 1$$
    $$x - 1 = 2y$$
    $$y = \frac{x - 1}{2}.$$

    We see that this is a function, so we define $F = f$.

4. $f(x) = 2x^2 + 1$
    The set of reals is understood to be the domain. Write f as a set of ordered pairs:

    $$f = \{(x, y) \mid y = 2x^2 + 1\}$$
    $$f^{-1} = \{(x, y) \mid x = 2y^2 + 1\}$$

    Solving for y:

    $$x = 2y^2 + 1$$
    $$x - 1 = 2y^2$$
    $$\frac{x - 1}{2} = y^2$$
    $$y = \pm \sqrt{\frac{x - 1}{2}}$$

    This is not a function, so we restrict the domain of f. Let $F = f$ where the domain of F is the set of nonnegative reals. That is,

    $$F = \{(x, y) \mid y = 2x^2 + 1, \ x \geq 0\}$$

$$F^{-1} = \{(x, y)\,|\,x = 2y^2 + 1,\ y \geq 0\}$$

Solving for y:

$$x = 2y^2 + 1$$

$$y = \pm\sqrt{\frac{x-1}{2}}$$

$$\uparrow$$

(*positive value since $y \geq 0$*)

$$y = \sqrt{\frac{x-1}{2}},\ \text{which is a function.}$$

One of the main topics of this course is the graphing of functions. Sometimes when graphing the inverse of a function, we can simplify our work by using a geometrical property of inverses. If (x, y) is a point on the graph of the original function, then (y, x) is a corresponding point on the graph of the inverse. Thus the inverse is symmetric with respect to the line $x = y$. That is, if we were to graph the function and then place a mirror along the line $x = y$, the mirror image would represent the graph of the inverse. It was for this reason that the line $x = y$ was drawn in Figure 1.24, 1.25, and 1.26, and, as you can see, in each figure the inverse is symmetric with respect to this line.

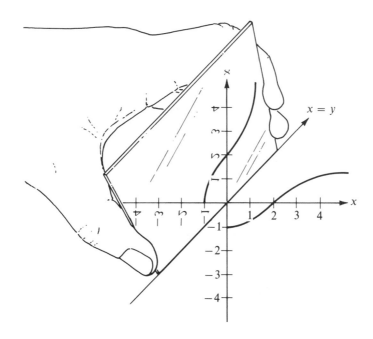

PROBLEM SET 1.5

A Problems

Suppose a mirror is held as illustrated in Problems 1–8. Draw the reflection of the curve in the mirror.

1.

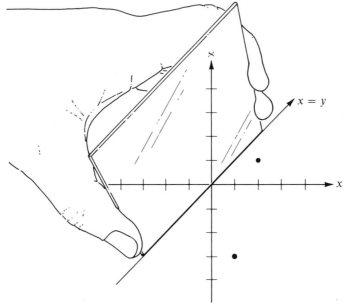

2.

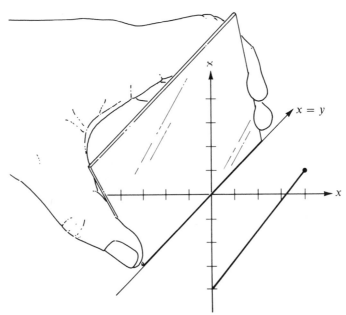

3.

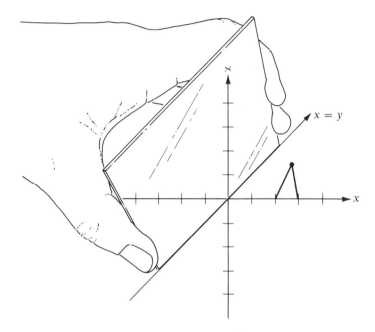

4.

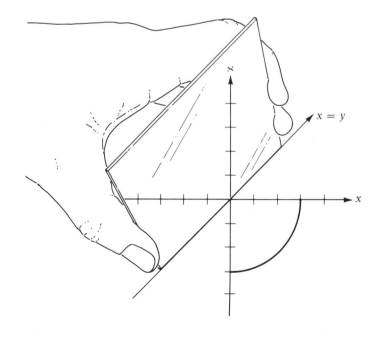

5.

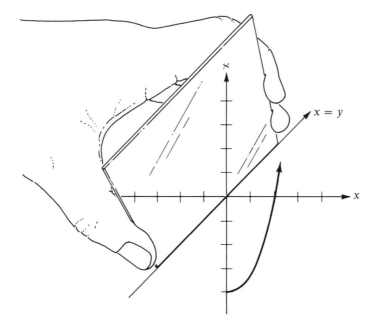

6.

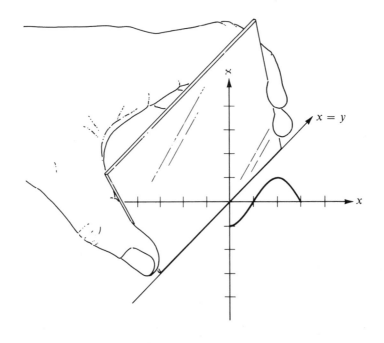

7.

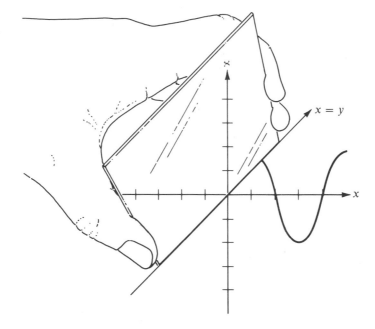

8.

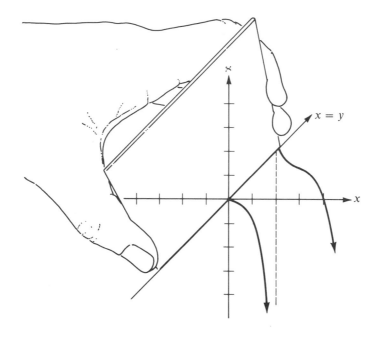

Determine which pairs of functions in Problems 9–18 are inverses.

EXAMPLE:

$$f(x) = \frac{2}{5}x + 3, \qquad g(x) = \frac{5}{2}x + 5$$

$$f[g(x)] = f\left(\frac{5}{2}x + 5\right)$$

$$= \frac{2}{5}\left(\frac{5}{2}x + 5\right) + 3$$

$$= x + 2 + 3$$

$$= x + 5$$

Since $f[g(x)] \neq x$, f and g are not inverses.

9. $f(x) = 5x + 3$; $g(x) = (x - 3)/5$ 10. $f(x) = \frac{2}{3}x + 2$; $g(x) = \frac{3}{2}x + 3$

11. $f(x) = \frac{4}{5}x + 4$; $g(x) = \frac{5}{4}x + 3$ 12. $f(x) = 1/x$: $x \neq 0$; $g(x) = 1/x$: $x \neq 0$

13. $f(x) = x^2$: $x < 0$; $g(x) = \sqrt{x}$: $x > 0$ 14. $f(x) = x^2$: $x \geq 0$; $g(x) = \sqrt{x}$: $x \geq 0$

B Problems

15. $f(x) = 2x^2 + 1$: $x \geq 0$; $g(x) = \frac{1}{2}\sqrt{2x - 2}$: $x \geq 1$

16. $f(x) = 2x^2 + 1$: $x \leq 0$; $g(x) = -\frac{1}{2}\sqrt{2x - 2}$: $x \geq 1$

17. $f(x) = (x + 1)^2$: $x \geq -1$; $g(x) = -1 - \sqrt{x}$: $x > 0$

18. $f(x) = (x + 1)^2$: $x \geq -1$; $g(x) = -1 + \sqrt{x}$: $x \geq 0$

Given the function f in Problems 19–28:
  a. *find the inverse of f;*
  b. *graph f and its inverse;*
  c. *determine if the inverse is a function;*
  d. *if the inverse is not a function, define a function F by limiting the domain of f so that F^{-1} is a function;*
  e. *graph F and F^{-1} if $F \neq f$.*

19. $f = \{(4, 5), (6, 3), (7, 1), (2, 4)\}$ 20. $f = \{(1, 4), (6, 1), (4, 5), (3, 4)\}$

21. $f = \{(x, y) | y = 2x + 3\}$ 22. $f = \{(x, y) | y = 3x + 1\}$

23. $f(x) = x + 5$ 24. $f(x) = \frac{1}{2}x + \frac{3}{2}$ 25. $f(x) = -\frac{1}{2}x^2$

26. $f(x) = 2x^2$ 27. $f(x) = |x|$ 28. $f(x) = |x + 1|$

If $y = f(x)$ is given by the graph in Figure 1.27, define a function $y = F(x)$ whose domain is $\{x | 0 \leq x \leq 7\}$ by limiting the domain of f so that F^{-1} exists. Then find the values requested in Problems 29–31.

29. a. $f(2)$ b. $f(4)$ c. $f(7)$ d. $f(10)$ e. $f(14)$

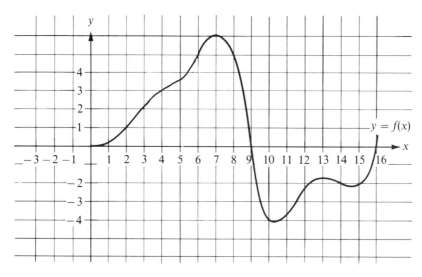

FIGURE 1.27   Graph of $y = f(x)$

30. a. $F(0)$ b. $F(4)$ c. $F(7)$ d. $f(16)$ e. $f(0)$

31. a. $F^{-1}(0)$ b. $F^{-1}(1)$ c. $F^{-1}(5)$ d. $F^{-1}(6)$ e. $F^{-1}(3)$

If $y = f(x)$ is given by the graph in Figure 1.28, define a function $y = F(x)$ whose domain is $\{x \,|\, 0 \leq x \leq 6\}$ by limiting the domain of f so that F^{-1} exists. Then find the values requested in Problems 32–35.

32. a. $f(2)$ b. $F(3)$ c. $F^{-1}(2)$ d. $F^{-1}(0)$

33. a. $f(3)$ b. $F(0)$ c. $F^{-1}(4)$ d. $F^{-1}(-3.5)$

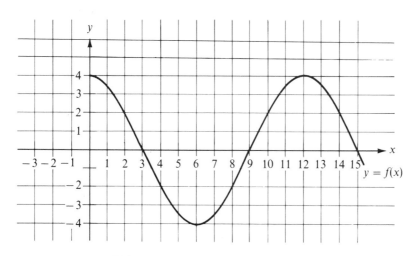

FIGURE 1.28   Graph of $y = f(x)$

34. a. $f(6)$ b. $f(12)$ c. $F^{-1}(-4)$ d. $F(5)$

35. a. $f(9)$ b. $f(15)$ c. $F(7)$ d. $F^{-1}(-2)$

Mind Bogglers

36. a. Graph the inverse of f for the function shown in Figure 1.27.
     b. Graph F^{-1} for the function shown in Figure 1.27.

37. a. Graph the inverse of f for the function shown in Figure 1.28.
     b. Graph F^{-1} for the function shown in Figure 1.28.

38. A function is said to be *increasing* on an interval I if $f(x_1) < f(x_2)$ whenever $x_1 < x_2$ and x_1 and x_2 are in I. Show that, if f is an increasing function, then f^{-1} is a function.

1.6 SUMMARY AND REVIEW

TERMS

Angle (1.2)
Arc (1.2)
Cartesian coordinate system (1.1)
Composition of functions (1.5)
Coterminal angles (1.2)
Degree (1.2)
Dependent variable (1.1)
Domain (1.1)
Even function (1.4)
Function (1.1)
Independent variable (1.1)
Initial side (1.2)
Inverse (1.5)
Line (1.3)
Linear function (1.3)
Minute (1.2)
Negative angle (1.2)
Odd function (1.4)
Ordered pair (1.1)

Origin (1.1)
Parabola (1.3)
Positive angle (1.2)
Quadrants (1.1)
Quadratic function (1.3)
Radian (1.2)
Range (1.1)
Rectangular coordinate system (1.1)
Relation (1.1)
Second (1.2)
Slope (1.3)
Standard-position angle (1.2)
Terminal side (1.2)
Translation (1.4)
Unit circle (1.2)
Vertex (1.2, 1.4)
x-axis (1.1)
y-axis (1.1)

CONCEPTS

(1.1) A **function** is a relation for which each member of the domain is associated with exactly one member of the range.

(1.1) **Functional notation**

(1.2) The **distance** between $P_1(x_1, y_1)$ and $P_2(x_2, y_2)$ is found by:

$$|P_1 P_2| = \sqrt{(x_2 - x_1)^2 + (y_2 - y_1)^2}.$$

(1.2) The **graph of an equation** or an **equation of a graph** means that there is a one-to-one correspondence between the set of all ordered pairs (x, y) that satisfy the equation and the set of all ordered pairs (x, y) that lie on the curve.

(1.2) **Measure** of angles:

$$\frac{\theta \text{ in degrees}}{360} \text{ is equivalent to } \frac{\theta \text{ in radians}}{2\pi}.$$

(1.2) **Equivalencies of angles:**

Degrees	0°	30°	45°	60°	90°	180°	270°	360°
Radians	0	$\pi/6$	$\pi/4$	$\pi/3$	$\pi/2$	π	$3\pi/2$	2π

(1.2) The **arc length**, s, cut by a central angle θ, measured in radians, of a circle of radius r is given by $s = r\theta$.

(1.3) **Linear and quadratic functions:**

Linear: $f(x) = mx + b$ or $y = mx + b$, where m is the slope and b is the y-intercept.

Quadratic: $f(x) = ax^2 + bx + c$ or $y = ax^2 + bx + c$, where $a \neq 0$ and c is the y-intercept.

(1.3, 1.4) Graphing linear and quadratic functions:

1. $y = mx$: graph passes through $(0, 0)$ with slope $= m =$ rise/run.
   $y = ax^2$: graph passes through $(0, 0)$; opens up if $a > 0$ and down if $a < 0$; is symmetric with respect to the y-axis.
2. $y - k = m(x - h)$: graph passes through (h, k) with slope m.
   $y - k = a(x - h)^2$: graph passes through (h, k); it is the graph of $y = x^2$ translated to the vertex (h, k) and symmetric with respect to the line $x = h$.

(1.5) The **inverse of a function** is found by interchanging the x and y components. Two functions f and g are inverses if and only if $f[g(x)] = g[f(x)] = x$. The graphs of a function and its inverse are symmetric with respect to the line $x = y$.

PROBLEM SET 1.6 *(Chapter 1 Test)*

1. (1.1) Let $f(x) = 4x - 3$ and $g(x) = (x + 3)/4$. Find:
   a. $f(3)$.   b. $g(9)$.   c. $f(w)$.   d. $g(4w - 3)$.   e. $f[g(x)]$.

2. (1.1, 1.2) Draw a unit circle. Draw an angle θ in standard position. The initial side intersects the unit circle at $(1, 0)$, and the terminal side intersects the unit circle at (a, b), as shown in Figure 1.29. Define functions $c(\theta) = a$ and $s(\theta) = b$. Now find:
   a. $c(90°)$.   b. $s(\pi/2)$.   c. $s(270°)$.   d. $c(\pi)$.   e. $s(2\pi)$.

3. (1.1, 1.2) Find the distance between $(3, f(3))$ and $(9, g(9))$ for the functions given in Problem 1.

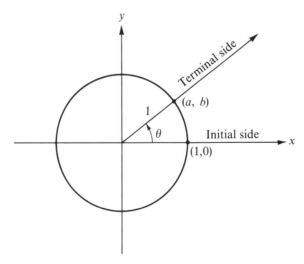

FIGURE 1.29

4. (1.2) From memory, fill in the blanks. ·

Angle in degrees	60°			300°	−210°			225°
Angle in radians		$\pi/2$	$5\pi/6$			$-11\pi/3$	3π	
Quadrant		—					—	

5. (1.2) A curve on a highway is laid out as the arc of a circle of radius 500 m. If the curve subtends a central angle of 18°, what is the distance around this section of road? Give the exact answer and an answer rounded off to the nearest meter.

6. (1.3, 1.4) Graph the following functions.
   a. $3x + 2y = 0$ b. $3x + 2y = 6$ c. $3x^2 + 2y = 0$ d. $3x^2 + 2y = 6$

7. (1.4) Graph the following functions.
   a. $y - 3 = \frac{2}{3}(x - 1)$ b. $y - 3 = \frac{2}{3}(x - 1)^2$
   c. $y = -2(x + 4) + 2$ d. $y = -2(x + 4)^2 + 2$

8. (1.4) Let $y = f(x)$ be the function given by Figure 1.30. Notice that the domain is $0 \le x \le 10$.
   a. What is the range?   b. Graph $y - \pi = f(x)$.
   c. Graph $y = f(x + \pi)$.   d. Graph $y + 4 = f(x - 3)$.

9. (1.5)
   a. Find the inverse of each function given in Problem 6.
   b. Which of the inverses in part (a) are functions? If the inverse is not a function, limit the domain to define a function F so that F^{-1} is also a function.

10. (1.5) Graph the inverse of f for the function given in Problem 8. Is it also a function?

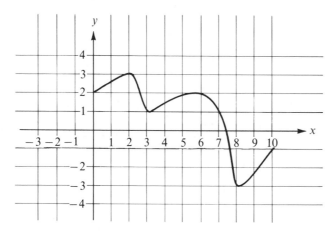

FIGURE 1.30   Graph of $y = f(x)$

In this chapter we define six functions that are related to an angle: the sine, cosine, tangent, cosecant, secant, and cotangent. These functions not only form the basis for study in this course but also are necessary for further study in mathematics and the sciences. Thus this chapter introduces some of the most important ideas you will learn in mathematics.

During the second half of the second century B.C., the astronomer Hipparchus of Nicaea (c. 180–c. 125 B.C.) tabulated the first trigonometric tables in 12 books. It was he who used the 360° circle of the Babylonians and thus introduced into trigonometry the angle measure we still use today. Hipparchus' work formed the foundation for Ptolemy's Mathematical Syntaxis, *the most significant early work in trigonometry. Ptolemy acknowledged the earlier contributions of Hipparchus, whom he described as "a labor-loving and truth-loving man." It should be pointed out that these early works do not use the ideas of trigonometric* ratios, *which we introduce in this chapter. Rather, they use trigonometric* lines, *which take the form of chords of a circle. For a discussion of this method, see Carl Boyer's* A History of Mathematics *(New York: Wiley, 1968), Chapter 10.*

2

TRIGONOMETRIC FUNCTIONS

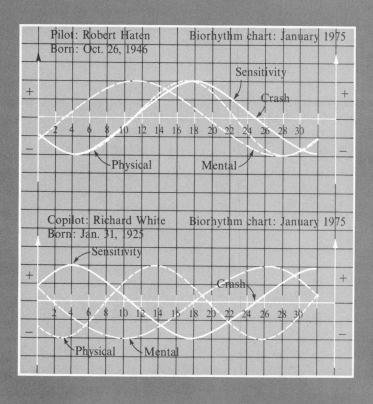

Pilot: Robert Haten   Biorhythm chart: January 1975
Born: Oct. 26, 1946

Sensitivity

Crash

2 4 6 8 10 12 14 16 18 20 22 24 26 28 30

Physical   Mental

Copilot: Richard White   Biorhythm chart: January 1975
Born: Jan. 31, 1925

Sensitivity

Crash

2 4 6 8 10 12 14 16 18 20 22 24 26 28 30

Physical   Mental

2.1 SINE AND COSINE *(You Don't Have to Be Religious to Study "the Sin")*

In the last chapter we considered the problem of drawing a unit circle with an angle θ in standard position and the intersection of the angle and the circle at the points $(1, 0)$ and (a, b), as shown in Figure 2.1. Then we defined functions c and s by $c(\theta) = a$ and $s(\theta) = b$. In this section we will generalize this idea. Suppose that $P(x, y)$ is a point on the terminal side a distance of r from the origin $(r \neq 0)$. If points A, B, and Q are labeled as shown in Figure 2.1, then $\triangle AOB \sim \triangle POQ$ (Problem 40, Section 1.3), and therefore the sides are proportional so that

$$a = \frac{a}{1} = \frac{x}{r} \quad \text{and} \quad b = \frac{b}{1} = \frac{y}{r}.$$

Notice that, since $\triangle POQ$ is a right triangle,

$$x^2 + y^2 = r^2$$

or

$$r = \sqrt{x^2 + y^2}$$

by the Pythagorean Theorem. Thus:

$$c(\theta) = a = \frac{x}{r} \quad \text{and} \quad s(\theta) = b = \frac{y}{r}.$$

This generalization allows us to find additional values for the functions c and s. For example, if we wish to find $c(45°)$, we can find $c(45°) = a$, which is the first component of the point (a, b) on the unit circle that is intersected by the terminal side of a $45°$ angle. (See Figure 2.2.) But instead of finding a, let's pick any point

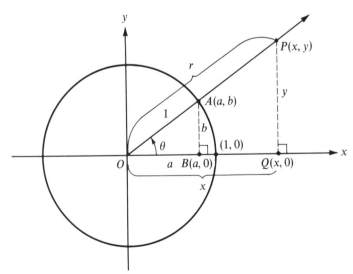

FIGURE 2.1

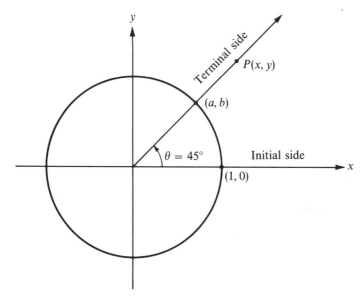

FIGURE 2.2

on the terminal side different from $(0, 0)$—say $(2, 2)$. Notice that an angle of $45°$ determines a ray that bisects the first quadrant. Since (x, y) represents points on the terminal side (x and y are both positive in the first quadrant), we see that $x = y$.

$$c(45°) = \frac{x}{r}$$

$$= \frac{x}{\sqrt{x^2 + y^2}}$$

$$= \frac{2}{\sqrt{4 + 4}}$$

$$= \frac{2}{2\sqrt{2}}$$

$$= \frac{1}{\sqrt{2}} \quad \text{or} \quad \frac{1}{2}\sqrt{2}$$

What if we choose $(1, 1)$ on the terminal side? Then

$$c(45°) = \frac{1}{\sqrt{1 + 1}}$$

$$= \frac{1}{2}\sqrt{2}.$$

Any point different from $(0, 0)$ on the terminal side gives the same result.

EXAMPLES:

1. Find $s(45°)$. By definition, $s(\theta) = b = y/r$, so we choose *any* point except $(0,0)$ on the terminal side—say $(1, 1)$—and find

$$s(45°) = \frac{y}{\sqrt{x^2 + y^2}}$$

$$= \frac{1}{\sqrt{1 + 1}}$$

$$= \frac{1}{\sqrt{2}} \quad \text{or} \quad \frac{1}{2}\sqrt{2}.$$

2. Find $c(5\pi/4)$. We have to find a point on the terminal side, as shown in Figure 2.3. Notice that the terminal side bisects the third quadrant and that x and y are both negative. Thus, $x = y$, so we choose *any* point except $(0,0)$ on the terminal side— say $(-1, -1)$.

$$c\left(\frac{5\pi}{4}\right) = \frac{x}{r}$$

$$= \frac{-1}{\sqrt{(-1)^2 + (-1)^2}}$$

$$= \frac{-1}{\sqrt{2}}.$$

$$= -\frac{1}{2}\sqrt{2}$$

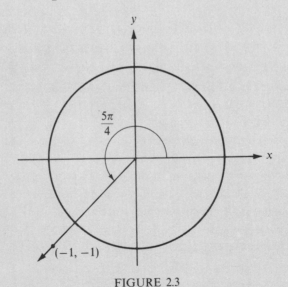

FIGURE 2.3

These functions c and s are called the *cosine* and *sine functions* in trigonometry and are usually written $c(\theta) = \cos\theta$ and $s(\theta) = \sin\theta$. There are many properties associated with these functions. For example, $\cos\theta$ will have the same value as $\cos\theta'$ if θ and θ' are coterminal, since the definition of cosine requires that we choose any point different from the vertex and on the terminal side and since, by definition, coterminal angles have the same terminal side. This point can be summarized by saying

$$\cos(\theta + 2\pi) = \cos\theta \quad \text{and} \quad \sin(\theta + 2\pi) = \sin\theta,$$

since $\theta + 2\pi$ and θ are coterminal angles. The following are also true.

$$\cos(\theta - 2\pi) = \cos\theta \qquad \sin(\theta - 2\pi) = \sin\theta$$
$$\cos(\theta + 4\pi) = \cos\theta \qquad \sin(\theta - 6\pi) = \sin\theta$$

And, more generally, we can say that

$$\cos(\theta \pm 2n\pi) = \cos\theta \quad \text{and} \quad \sin(\theta \pm 2n\pi) = \sin\theta$$

for any integer n. In other words, the functional values of the cosine and the sine functions repeat themselves after $\pm 2\pi$ ($\pm 360°$) units. We say that these functions are *periodic* with period 2π.

EXAMPLES:

1. Find $\cos 405°$. Since $405°$ is coterminal with $45°$ ($405° = 360° + 45°$), we have $\cos 405° = \cos 45° = \frac{1}{2}\sqrt{2}$.

2. Find $\sin(-7\pi/4)$. Since $-7\pi/4$ is coterminal with $\pi/4$, we have $\sin(-7\pi/4) = \sin\pi/4 = \frac{1}{2}\sqrt{2}$.

3. Find $\cos 60°$. We need to find a point (x, y) on the terminal side of this angle, as shown in Figure 2.4. We know from geometry that in a $30°$-$60°$ right triangle the length of the hypotenuse is twice the length of the shorter leg and the longer leg is $\sqrt{3}$ times the length of the shorter leg. Thus, $r = 2x$ and

$$\cos 60° = \frac{x}{r}$$
$$= \frac{x}{2x}$$
$$= \frac{1}{2}.$$

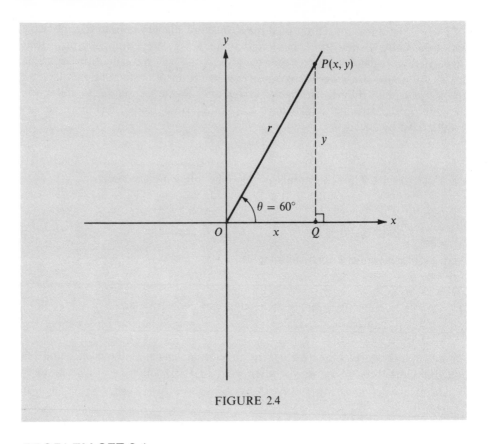

FIGURE 2.4

PROBLEM SET 2.1

A Problems

Each of the angles in Problems 1–6 is coterminal with an angle θ so that $0 \le \theta \le 90°$. Find θ using the unit of measurement (radians or degrees) given in the problem.

1. $-300°$ 2. $-330°$ 3. $17\pi/4$ 4. -6π 5. $9\pi/2$ 6. $-675°$

Find the functional values in Problems 7–26.

7. $\sin 0°$ 8. $\sin 30°$ 9. $\sin \pi/4$ 10. $\sin 60°$ 11. $\sin \pi/2$

12. $\sin 180°$ 13. $\sin 270°$ 14. $\cos 0$ 15. $\cos \pi/6$ 16. $\cos 45°$

17. $\cos \pi/2$ 18. $\cos 90°$ 19. $\sin \pi$ 20. $\cos 3\pi/2$ 21. $\cos(-300°)$

22. $\sin 390°$ 23. $\sin 17\pi/4$ 24. $\cos(-6\pi)$ 25. $\cos 9\pi/2$ 26. $\sin(-675°)$

Simplify the expressions in Problems 27–41. Recall that $\sqrt{x^2} = |x|$. This means that $\sqrt{x^2} = x$ if x is positive and $-x$ if x is negative.

27. $\sqrt{2x^2}$ 28. $\sqrt{2x^2}$ if x is positive

29. $\sqrt{2x^2}$ if x is negative

30. $x/\sqrt{2x^2}$ if x is negative

31. $x/\sqrt{4x^2}$ if x is positive

32. $\sin 30° + \cos 0$

33. $2 \cos \pi/2$

34. $\cos 2(\pi/2)$

35. $\sin 2(\pi/4)$

36. $2 \sin \pi/4$

37. $\sin \pi/2 + 3 \cos \pi/2$

38. $\cos^2 \pi/4$ (Note: This means $[\cos \pi/4]^2$. We will write $\cos^2 \theta = [\cos \theta]^2$ and $\sin^2 \theta = [\sin \theta]^2$.)

39. $\sin^2 60°$

40. $\sin^2 \pi/2 + \cos^2 \pi/2$

41. $\sin^2 \pi/6 + \cos^2 \pi/2$

B Problems

42. In this section we found $\cos 45°$ by choosing particular points other than the vertex on the terminal side. If we let $x = y$, we can find $\cos 45°$ in general:

$$\cos 45° = \frac{x}{\sqrt{x^2 + y^2}}$$

$$= \frac{x}{\sqrt{x^2 + x^2}}$$

$$= \frac{x}{\sqrt{2x^2}}$$

$$= \frac{x}{x\sqrt{2}}$$

$$= \frac{1}{\sqrt{2}} \quad \text{or} \quad \frac{1}{2}\sqrt{2}.$$

Now, from Problem 27, $\sqrt{2x^2} \neq x\sqrt{2}$. Why is it that we can say $\sqrt{2x^2} = x\sqrt{2}$ for this problem?

43. In this section we found $\cos 5\pi/4$ by choosing a particular point on the terminal side. We noticed that the terminal side bisects the third quadrant and that x and y are both negative. Show that $\cos 5\pi/4 = -\frac{1}{2}\sqrt{2}$ by working with an arbitrary point (x, y)—not $(0, 0)$—as shown in Problem 42.

Find the values in Problems 44–47 by working with an arbitrary point (x, y)—not $(0, 0)$—as shown in Problem 42.

44. $\cos 135°$ 45. $\sin(-\pi/4)$ 46. $\sin 210°$ 47. $\cos 210°$

Mind Bogglers

48. What is the smaller angle between the hands of a clock at 12:25 P.M.?

49. a. Let $P(x, y)$ be any point in the plane. Show that $P(r \cos \theta, r \sin \theta)$ is a representation for P where θ is the standard-position angle formed by drawing ray $\overrightarrow{OP}$.

b. Let $A(\cos \alpha, \sin \alpha)$ and $B(\cos \beta, \sin \beta)$ be any two points on a unit circle. Use the distance formula to show that

$$|AB| = \sqrt{2 - 2(\cos \alpha \cos \beta + \sin \alpha \sin \beta)}.$$

2.2 TANGENT AND COFUNCTIONS
(Please Say Tan q)

In our discussion of sines and cosines in the last section, we considered certain ratios formed by choosing a point (x, y) a distance of $r > 0$ units from the origin on the terminal side of an angle θ. That is, $\cos \theta = x/r$ and $\sin \theta = y/r$ were defined. However, there are four other ratios to consider—namely, r/x, r/y, y/x, and x/y. In trigonometry each of these ratios has a name. As we saw in the last section, these ratios are functions of the angle θ and do not depend on the choice of the point P on the terminal side. The following definitions form the basis for this course and should be memorized before we continue.

Definition of the Trigonometric Functions: Let θ be an angle in standard position with a point $P(x, y)$ on the terminal side a distance of $r = \sqrt{x^2 + y^2}$ from the origin $(r \neq 0)$. Then the six trigonometric functions are defined as follows.

cosine function: $\cos \theta = \dfrac{x}{r}$ **secant** function: $\sec \theta = \dfrac{r}{x} \ (x \neq 0)$

sine function: $\sin \theta = \dfrac{y}{r}$ **cosecant** function: $\csc \theta = \dfrac{r}{y} \ (y \neq 0)$

tangent function: $\tan \theta = \dfrac{y}{x} \ (x \neq 0)$ **cotangent** function: $\cot \theta = \dfrac{x}{y} \ (y \neq 0)$

Using this definition, we can now find various functional values for θ.

EXAMPLES:

1. Find $\tan 45°$. For a $45°$ angle we know that $x = y$, so, by the definition of tangent, $\tan 45° = y/x = x/x = 1$.

2. Find $\sec 60°$. For a $60°$ angle the length of the hypotenuse (r) is twice the length of the shorter leg. Thus, $r = 2x$ and $\sec 60° = r/x = 2x/x = 2$.

3. Find $\tan 90°$. For a $90°$ angle, $x = 0$, and, from the definition, $\tan \theta = y/x \ (x \neq 0)$, so the tangent is *not defined* for $90°$. Notice also from the definition that $\sec 90°$ is not defined. For what angles are the cosecant and cotangent not defined?

By considering the examples of Sections 2.1 and 2.2, you should be able to verify the entries in the following table.

function \ angle θ	0	$\dfrac{\pi}{6}$	$\dfrac{\pi}{4}$	$\dfrac{\pi}{3}$	$\dfrac{\pi}{2}$	π	$\dfrac{3\pi}{2}$
$\cos \theta$	1	$\dfrac{\sqrt{3}}{2}$	$\dfrac{\sqrt{2}}{2}$	$\dfrac{1}{2}$	0	-1	0
$\sin \theta$	0	$\dfrac{1}{2}$	$\dfrac{\sqrt{2}}{2}$	$\dfrac{\sqrt{3}}{2}$	1	0	-1
$\tan \theta$	0	$\dfrac{\sqrt{3}}{3}$	1	$\sqrt{3}$	undef.	0	undef.
$\sec \theta$	1	$\dfrac{2}{\sqrt{3}}$	$\dfrac{2}{\sqrt{2}}$	2	undef.	-1	undef.
$\csc \theta$	undef.	2	$\dfrac{2}{\sqrt{2}}$	$\dfrac{2}{\sqrt{3}}$	1	undef.	-1
$\cot \theta$	undef.	$\dfrac{3}{\sqrt{3}}$	1	$\dfrac{\sqrt{3}}{3}$	0	undef.	0

This table of values is used extensively, and you should memorize it as you did multiplication tables in elementary school. There are, however, some hints that will make it easier to learn.

From the definition, $\cos \theta = x/r$ and $\sin \theta = y/r$, so

$$\cos^2 \theta + \sin^2 \theta = \frac{x^2}{r^2} + \frac{y^2}{r^2}$$

$$= \frac{x^2 + y^2}{r^2}.$$

But $r^2 = x^2 + y^2$, so

$$\cos^2 \theta + \sin^2 \theta = \frac{r^2}{r^2}.$$

$$\cos^2 \theta + \sin^2 \theta = 1$$

This equation restricts the values of $\cos \theta$ and $\sin \theta$: $\cos^2 \theta \le 1$ and $\sin^2 \theta \le 1$ (why?). Thus *both* sine and cosine have functional values between -1 and $+1$, inclusive:

$$-1 \le \cos \theta \le 1 \quad \text{and} \quad -1 \le \sin \theta \le 1.$$

Also notice the following pattern from the table:

	0	$\pi/6$	$\pi/4$	$\pi/3$	$\pi/2$	
$\cos\theta$	1	$\dfrac{\sqrt{3}}{2}$	$\dfrac{\sqrt{2}}{2}$	$\dfrac{1}{2}$	0	⟵ simplified form
$\cos\theta$	$\dfrac{\sqrt{4}}{2}$	$\dfrac{\sqrt{3}}{2}$	$\dfrac{\sqrt{2}}{2}$	$\dfrac{\sqrt{1}}{2}$	$\dfrac{\sqrt{0}}{2}$	⟵ easy-to-remember form

equal values

And the sine is simply the reversal of this pattern:

	0	$\pi/6$	$\pi/4$	$\pi/3$	$\pi/2$	
$\sin\theta$	0	$\dfrac{1}{2}$	$\dfrac{\sqrt{2}}{2}$	$\dfrac{\sqrt{3}}{2}$	1	⟵ simplified form
$\sin\theta$	$\dfrac{\sqrt{0}}{2}$	$\dfrac{\sqrt{1}}{2}$	$\dfrac{\sqrt{2}}{2}$	$\dfrac{\sqrt{3}}{2}$	$\dfrac{\sqrt{4}}{2}$	⟵ easy-to-remember form

For the tangent we use the following pattern:

	0	$\pi/6$	$\pi/4$	$\pi/3$	$\pi/2$	
$\tan\theta$	0	$\dfrac{\sqrt{3}}{3}$	1	$\sqrt{3}$	undef.	⟵ simplified form
$\tan\theta$	$\sqrt{\dfrac{0}{4}}$	$\sqrt{\dfrac{1}{3}}$	$\sqrt{\dfrac{2}{2}}$	$\sqrt{\dfrac{3}{1}}$	$\sqrt{\dfrac{4}{0}}$	easy-to-remember form ⟵ (Division by 0 means undefined.)

Another way to remember the values of the tangent is to notice that

$$\frac{\sin\theta}{\cos\theta} = \frac{\dfrac{y}{r}}{\dfrac{x}{r}} \qquad (\cos\theta \neq 0)$$

$$= \frac{y}{r}\cdot\frac{r}{x} \qquad (x \neq 0)$$

$$= \frac{y}{x} \qquad (x \neq 0),$$

which is the definition of tangent. Therefore:

$$\tan\theta = \frac{\sin\theta}{\cos\theta} \qquad (\cos\theta \neq 0).$$

The rest of the table is easy to remember; the definition of the secant says that it is the reciprocal of the cosine. This means that $\sec \theta = 1/\cos \theta$, since

$$\frac{1}{\cos \theta} = \frac{1}{x/r} \qquad \text{(from the definition of cosine)}$$

$$= 1 \cdot \frac{r}{x} \qquad (x \neq 0)$$

$$= \frac{r}{x} \qquad \text{(the definition of secant)}.$$

Similar results hold for the cosecant and cotangent:

$$\sec \theta = 1/\cos \theta \qquad (\cos \theta \neq 0)$$
$$\csc \theta = 1/\sin \theta \qquad (\sin \theta \neq 0)$$
$$\cot \theta = 1/\tan \theta \qquad (\tan \theta \neq 0)$$

When using these definitions to remember the table, you must keep in mind that the reciprocal of 0 is undefined. (This, of course, is not true in general, but it will help you to remember the entries in the table.)

Don't confuse these reciprocal relationships with the idea of cofunctions. The sine and cosine are cofunctions, as are the secant and cosecant, and the tangent and cotangent.

For general angles, particularly those in quadrants II, III, and IV, we will now define the notion of reference angle. Consider four examples of an angle, one in each of the four quadrants.

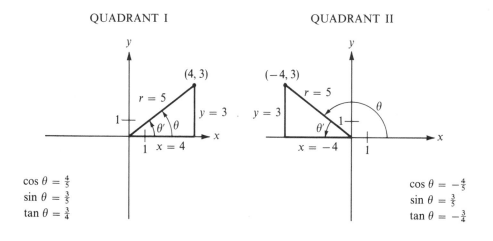

QUADRANT I

$\cos \theta = \frac{4}{5}$
$\sin \theta = \frac{3}{5}$
$\tan \theta = \frac{3}{4}$

QUADRANT II

$\cos \theta = -\frac{4}{5}$
$\sin \theta = \frac{3}{5}$
$\tan \theta = -\frac{3}{4}$

QUADRANT III   QUADRANT IV

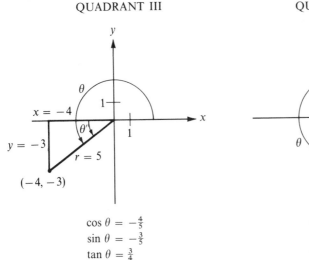
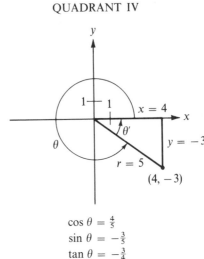

$$\cos \theta = -\tfrac{4}{5}$$
$$\sin \theta = -\tfrac{3}{5}$$
$$\tan \theta = \tfrac{3}{4}$$

$$\cos \theta = \tfrac{4}{5}$$
$$\sin \theta = -\tfrac{3}{5}$$
$$\tan \theta = -\tfrac{3}{4}$$

Given an angle θ, we define its *reference angle* θ' to be the smallest positive acute angle the terminal side makes with the x-axis. Notice that the reference angle θ' and angle θ are the same in the first quadrant and that the reference angle for each of the above examples is the same. The only change in the answers is in the sign of the result. In other words, if f represents a trigonometric function, then

$$f(\theta) = \pm f(\theta')$$
└─ where this sign depends on the
quadrant and the function.

This relationship is called the *reduction principle.*

We determine the signs from the definition of the functions.

	Quadrant I	Quadrant II	Quadrant III	Quadrant IV
	x pos y pos r pos	x neg y pos r pos	x neg y neg r pos	x pos y neg r pos
$\cos \theta = x/r$	pos	neg	neg	pos
$\sin \theta = y/r$	pos	pos	neg	neg
$\tan \theta = y/x$	pos	neg	pos	neg
Summary:	All positive	Sine positive	Tangent pos	Cosine positive

This table can be summarized by remembering the following.

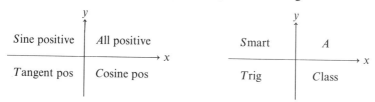

Sine positive	All positive		Smart	A
Tangent pos	Cosine pos		Trig	Class

Signs of Trig Functions   Easy-to-Remember Form: *A Smart Trig Class*

The values of sec θ, csc θ, and cot θ are the reciprocals of the above values.

EXAMPLES:

1. $\tan 210° = \underset{\underset{\substack{\text{Quadrant III:} \\ \text{tangent pos,} \\ \text{secant neg}}}{\uparrow}}{+} \qquad \underset{\underset{\text{reference angle}}{\uparrow}}{\tan 30°} = \underset{\underset{\text{from memorized table}}{\uparrow}}{\dfrac{\sqrt{3}}{3}}$

$\qquad\qquad\qquad\quad \downarrow \qquad\qquad\qquad \downarrow \qquad\qquad\qquad \downarrow$

2. $\sec 210° = \qquad - \qquad \sec 30° = -\dfrac{2}{\sqrt{3}} \text{ or } -\dfrac{2}{3}\sqrt{3}$

3. $\csc 3\pi/2 = -1$ (If it is not in any quadrant, then it comes directly from the memorized table.)

4. $\sec(-5\pi/3) = +\sec \pi/3 = 2$ (Sketch the angle if necessary to find the quadrant and the reference angle.)

5. $\cot(-7\pi/6) = -\cot \pi/6 = -\sqrt{3}$

PROBLEM SET 2.2

A Problems

In Problems 1–6 give the functional value from memory.

1. a. $\tan \pi/4$ b. $\cos 0$ c. $\sin 60°$ d. $\sec \pi/6$ e. $\csc 0$ f. $\tan \pi/6$

2. a. $\cos 270°$ b. $\cos 30°$ c. $\sec \pi/4$ d. $\tan 180°$ e. $\sin 45°$ f. $\sin \pi$

3. a. $\sin \pi/2$ b. $\sec 0°$ c. $\tan 0$ d. $\sec \pi/6$ e. $\cos \pi/4$ f. $\cot 90°$

4. a. $\cot \pi$ b. $\sec \pi/4$ c. $\tan 90°$ d. $\tan 60°$ e. $\cos \pi/3$ f. $\sec \pi/3$

5. a. $\cot 45°$ b. $\cos \pi$ c. $\sin 3\pi/2$ d. $\sin 0°$ e. $\sec \pi/2$ f. $\sec 270°$

6. a. $\csc \pi/2$ b. $\cos 90°$ c. $\sec \pi$ d. $\tan 270°$ e. $\sin \pi/6$ f. $\csc 3\pi/2$

In Problems 7–16 write the answer in simplest form.

7. $\sin \pi/6 \csc \pi/6$

8. $\csc \pi/2 \sin \pi/2$

9. $\sin^2 \pi/3 + \cos^2 \pi/3$

10. $\sin^2 \pi/6 + \cos^2 \pi/3$

11. a. $\cos(\pi/4 - \pi/2)$ b. $\cos \pi/4 - \cos \pi/2$

12. a. $\tan(2 \cdot 30°)$ b. $2 \tan 30°$

13. a. $\csc\left(\dfrac{1}{2} \cdot 60°\right)$ b. $\dfrac{\csc 60°}{2}$

14. a. $\cos(\pi/2 - \pi/6)$ b. $\cos \pi/2 \cos \pi/6 + \sin \pi/2 \sin \pi/6$

15. a. $\tan(2 \cdot 60°)$ b. $\dfrac{2 \tan 60°}{1 - \tan^2 60°}$

16. a. $\cos\left(\dfrac{1}{2} \cdot 60°\right)$ b. $\sqrt{\dfrac{1 + \cos 60°}{2}}$

B Problems

Find the values of the six trigonometric functions for an angle θ in standard position whose terminal side passes through the points in Problems 17–28. Draw a picture showing θ and the reference angle θ′.

17. (3, 4)	18. (−3, 4)	19. (−3, −4)	20. (3, −4)	21. (5, 12)
22. (−5, −12)	23. (5, −12)	24. (−5, 12)	25. (2, −5)	26. (−6, 1)
27. (−4, −5)	28. (−5, 2)			

Find the values of the six trig functions for each of the angles given in Problems 29–40.

29. 450°	30. −5π/4	31. −π/6	32. 120°	33. −2π
34. 7π/6	35. 390°	36. −120°	37. −135°	38. 135°
39. 3π	40. 8π/3			

Although we've been limiting ourselves to angles whose reference angles are 0°, 30°, 45°, 60°, or 90°, we can estimate the functional values for other angles. From the definitions of the trigonometric functions, estimate to one decimal place the numbers in Problems 41–51.

EXAMPLE: For $\tan 110°$ we use a unit circle, as shown in Figure 2.5. Draw the terminal side at 110°. Estimate $y \approx 0.92$ and $x \approx -0.35$. Then, by the definition, $\tan 110° \approx 0.92/(-0.35) \approx -2.7$ to the nearest tenth.

41. $\cos 50°$	42. $\sec 70°$	43. $\csc 150°$	44. $\cot 250°$
45. $\tan(-20°)$	46. $\csc 190°$	47. $\sec(-190°)$	48. $\sin 20°$
49. $\sin 320°$	50. $\tan 80°$	51. $\cot(-100°)$	

VALUES OF TRIGONOMETRIC FUNCTIONS BY CALCULATOR

Many calculators have built-in trig functions. However, attention must be given to the unit of measure used. For a calculator with a degree-radian switch, simply depress the given angle (in degrees or radians) and then the appropriate trig-function key.

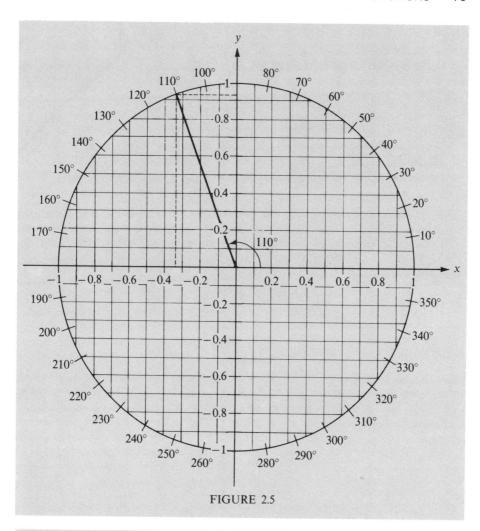

FIGURE 2.5

EXAMPLES:

1. sin 50°: Switch key to degrees; depress the following keys:

    | 5 | | 0 | | sin |

    The answer 0.7660444431 is now displayed.

2. sin π/12: Switch key to radians; depress the following keys:

    | π | | ÷ | | 1 | | 2 | | = | | sin |

    The answer 0.2588190451 is now displayed.

3. tan 70° 23′ 40″: Switch key to degrees; this measure must first be changed to a decimal-degree measure, as shown in Section 1.2:

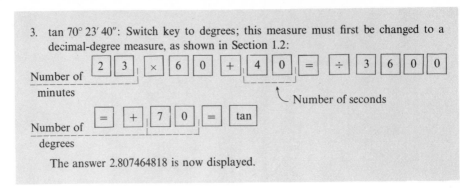

The answer 2.807464818 is now displayed.

Other calculators are preprogrammed for degrees only. For these machines you must first convert radians to degrees (as shown in Section 1.2) by using the formula (angle in degrees) = $180/\pi$ (angle in radians).

52. *Calculator Problem.* Find the values of the numbers in Problems 41–46 correct to six decimal places.

53. *Calculator Problem.* Find the values of the numbers in Problems 47–51 correct to six decimal places.

VALUES OF TRIGONOMETRIC FUNCTIONS BY TABLE

Sometimes we require more accuracy than can be obtained by the methods of Problems 41–51 but have no access to a calculator. Mathematicians have developed tables to find very precise trig values. Table III in the back of this book gives the values of the trigonometric functions of angles from 0° to 90°, and Table IV gives the values for the sine, tangent, cotangent, and cosine measured in radians from 0 to 2.

For angles between 0° and 45° we first find the angle to the nearest tenth in the left-hand column headed *Deg.* Next, we read across that row to find the value of the desired function named at the *top.* For example, cos 22.1° (from Table III) is found as follows.

Deg	Sin	Tan	Cot	Cos
18.0				
↓				
Read down this column until the desired angle is found.				
↓				
(22.1)		Read across to find cos 22.1.		(0.9265)

For angles between 45° and 90°, we find the desired angle in the *right-hand* column. Then we read across that row to find the value of the desired function named at the *bottom.* For example, cot 57.6° (from Table III) is found as follows.

Disregard the headings at the top when seeking values for angles between 45° and 90°.

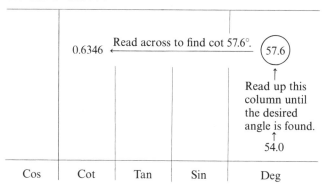

Cos	Cot	Tan	Sin	Deg

EXAMPLES: Verify the following from Table III: (1) sin 34.4° = 0.5650; (2) cos 54.2° = 0.5850; (3) tan 70.2° = 2.778; (4) cot 46.7° = 0.9424.

Since all reference angles are between 0° and 90°, Table III can be used to approximate any of the trig values.

54. Find the values of the numbers in Problems 41–46 correct to four decimal places by using Table III.

55. Find the values of the numbers in Problems 47–51 correct to four decimal places by using Table III.

In Problems 56–64 use Table III or IV or a calculator to find the functional value. Remember that, if degrees are not stated, the units are in radians.

56. tan 56.2°   57. cot 78.4°   58. sin 1°   59. sin 1   60. tan(−1.5)

61. cos(−0.48)   62. cos 48.3°   63. sin 21.48°   64. cos(−65.21°)

Note: For Problems 63 and 64, approximate the angle to the nearest tenth and use Table III, use interpolation, or use a calculator.

65. a. If θ is in Quadrant I, then $\theta + \pi$ is in Quadrant III with a reference angle θ. Use this fact and the reduction principle to show that $\sin(\theta + \pi) = -\sin \theta$ if θ is in Quadrant I.
  b. Show that $\sin(\theta + \pi) = -\sin \theta$ if θ is in Quadrant II.
  c. Show that $\sin(\theta + \pi) = -\sin \theta$ if θ is in Quadrant III.
  d. Show that $\sin(\theta + \pi) = -\sin \theta$ if θ is in Quadrant IV.
  e. By considering parts (a)–(d), show that $\sin(\theta + \pi) = -\sin \theta$ for any angle θ.

66. Show that $\cos(\theta + \pi) = -\cos \theta$ for any angle θ. (Hint: see Problem 65.)

Mind Bogglers

67. *Calculator Problem.* In more advanced mathematics courses it is shown that

$$\sin x = x - \frac{x^3}{3!} + \frac{x^5}{5!} - \frac{x^7}{7!} + \cdots$$

where $n! = n(n-1)(n-2)\ldots 3 \cdot 2 \cdot 1$. Find sin 1 correct to four decimal places by using this equation. (Remember that the 1 in sin 1 refers to radian measure.)

68. *Calculator Problem.* It is known that

$$\cos x = 1 - \frac{x^2}{2!} + \frac{x^4}{4!} - \frac{x^6}{6!} + \cdots.$$

Use this equation to find cos 1 correct to four decimal places.

69. *Computer Problem.* If you have access to a computer, write a program that will output a table of trig values for the sine, cosine, and tangent for every degree from 0° to 45°.

2.3 GRAPHS OF SINE, COSINE, AND TANGENT
(Up-Down, Down-Up, or Up-Up and Away)

Often the nature of a function can be understood more easily if we look at a picture or a graph of it. With most functions we begin by plotting points to determine the general shape. We then generalize so we can graph the curve without a lot of calculations concerning points.

SINE CURVE *(Up-down)*

To graph $y = \sin x$, we can begin by plotting familiar values for the sine:

$x = $ real number	0	$\pi/6$	$\pi/4$	$\pi/3$	$\pi/2$	π	$3\pi/2$
$y = \sin x$	0	1/2	$\sqrt{2}/2$	$\sqrt{3}/2$	1	0	-1
y approx.	0	0.5	0.71	0.87	1	0	-1

The difficulty with this method is that approximate values must be plotted. We can help matters a little by setting up a scale on the x-axis that is in units of π (we've chosen 12 squares $= \pi$ units in Figure 2.6). We can plot additional values using the reduction principle of the previous section. (By the way, we could also use Table III or IV or a calculator to generate a table of values.)

$x = $ real number	$2\pi/3$	$3\pi/4$	$5\pi/6$	$7\pi/6$	$5\pi/4$	$4\pi/3$	$5\pi/3$	$7\pi/4$	$11\pi/6$
Quadrant; sign of $\sin x$	II; +	II; +	II; +	III; −	III; −	III; −	IV; −	IV; −	IV; −
Reference angle	$\pi/3$	$\pi/4$	$\pi/6$	$\pi/6$	$\pi/4$	$\pi/3$	$\pi/3$	$\pi/4$	$\pi/6$
$y = \sin x$	$\sqrt{3}/2$	$\sqrt{2}/2$	1/2	$-1/2$	$-\sqrt{2}/2$	$-\sqrt{3}/3$	$-\sqrt{3}/2$	$-\sqrt{2}/2$	$-1/2$
y approx.	0.87	0.71	0.5	-0.5	-0.71	-0.87	-0.87	-0.71	-0.5

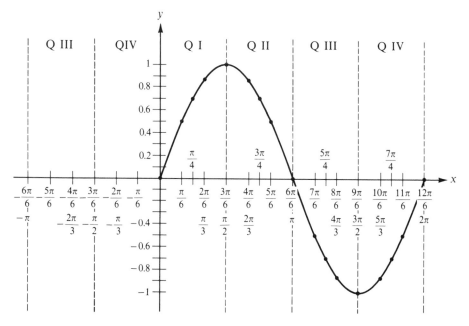

FIGURE 2.6   Graph of $y = \sin x$ for $0 \le x \le 2\pi$.

Notice that, when x is in Quadrant I, then $0 < x < \pi/2$, which does *not* correspond to the first quadrant of the graph $y = \sin x$. In Figure 2.6 we've shown the intervals corresponding to the quadrants of the angle x. The points from the preceding table are plotted in Figure 2.6 and are connected by a smooth curve called the *sine curve*.

What about values other than $0 \le x \le 2\pi$? Since the sine function is periodic, with period 2π, we see that angles greater than 2π or less than 0 simply repeat the values already plotted. The entire sine curve continues as indicated in Figure 2.7.

Notice that, for the base period shown, the sine curve starts at $(0, 0)$, goes *up* to $(\pi/2, 1)$ and then *down* to $(3\pi/2, -1)$ passing through $(\pi, 0)$, and then goes back up to $(2\pi, 0)$, which completes one period. We can summarize a technique for sketching the sine curve (see Figure 2.8):

1. Plot the endpoints of the base period—namely $(0, 0)$ and $(2\pi, 0)$.
2. Plot the midpoint, $(\pi, 0)$.
3. Halfway between these points, plot the highest (up) point, $(\pi/2, 1)$, and the lowest (down) point, $(3\pi/2, -1)$.
4. Now the sine curve is "framed."

COSINE CURVE *(Down-up)*

We can graph the cosine curve by plotting points, as we did with the sine curve. We'll leave the details of plotting these points as an exercise and summarize the results by "framing" the cosine curve (see Figure 2.9):

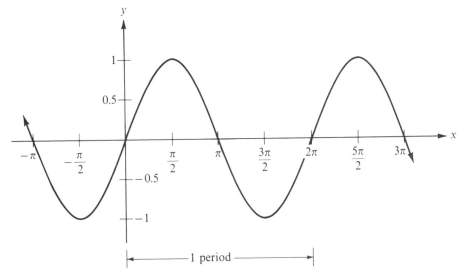

FIGURE 2.7   Graph of $y = \sin x$.

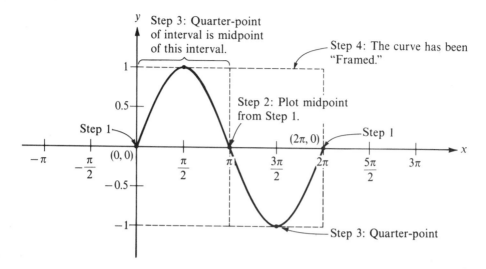

FIGURE 2.8   Framing the sine curve.

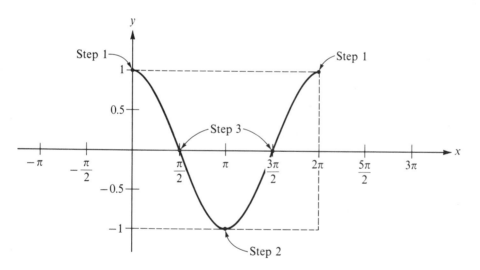

FIGURE 2.9   Framing the cosine curve.

1. Plot the endpoints of the base period—namely $(0, 1)$ and $(2\pi, 1)$.
2. Plot the midpoint, $(\pi, -1)$.
3. Halfway between these points, plot the points $(\pi/2, 0)$ and $(3\pi/2, 0)$.
4. Now the cosine curve is "framed."

Since values for x greater than 2π or less than 0 are coterminal with those already considered, we indicate the entire cosine curve in Figure 2.10.

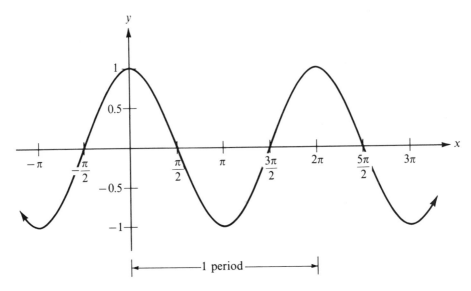

FIGURE 2.10   Graph of $y = \cos x$.

TANGENT CURVE *(Up-up and Away)*

By setting up a table of values and plotting points (the details are left as an exercise), we notice that $y = \tan x$ doesn't exist at $\pi/2$, $3\pi/2$, or $\pi/2 \pm n\pi$ for any integer n. The lines $x = \pi/2$, $x = 3\pi/2, \ldots$, $x = \pi/2 \pm n\pi$ for which the tangent is not defined are called *asymptotes*.

We now "frame" the tangent curve as follows (see Figure 2.11):

1. Plot the center point, $(0, 0)$.
2. Draw a pair of adjacent asymptotes—say $-\pi/2$ and $\pi/2$.
3. Halfway between the center point and the asymptotes, plot the points $(\pi/4, 1)$ and $(-\pi/4, -1)$.
4. The tangent curve is now "framed."

The entire tangent curve is indicated in Figure 2.12. Even though the curve repeats for values of x greater than 2π or less than 0, it also repeats after it has passed through an interval with length π. This result can be shown algebraically if we use the answers to Problems 65 and 66 of Section 2.2. Since $\sin(\theta + \pi) = -\sin\theta$ and $\cos(\theta + \pi) = -\cos\theta$, we have

$$\tan(\theta + \pi) = \frac{\sin(\theta + \pi)}{\cos(\theta + \pi)} = \frac{-\sin\theta}{-\cos\theta} = \frac{\sin\theta}{\cos\theta} = \tan\theta.$$

Since $\tan(\theta + \pi) = \tan\theta$, then $\tan(\theta + n\pi) = \tan\theta$ for any integer n, and we see that the tangent has a period of π.

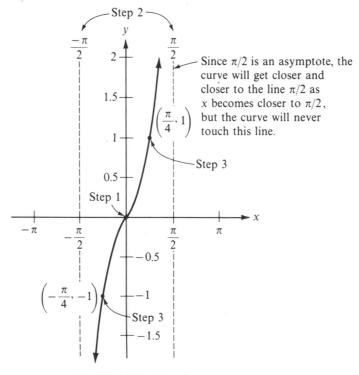

Step 2

$-\dfrac{\pi}{2}$ $\dfrac{\pi}{2}$

Since $\pi/2$ is an asymptote, the
curve will get closer and
closer to the line $\pi/2$ as
x becomes closer to $\pi/2$,
$\left(\dfrac{\pi}{4}, 1\right)$ but the curve will never
touch this line.

Step 3

Step 1

$\left(-\dfrac{\pi}{4}, -1\right)$

Step 3

FIGURE 2.11   Framing the tangent curve.

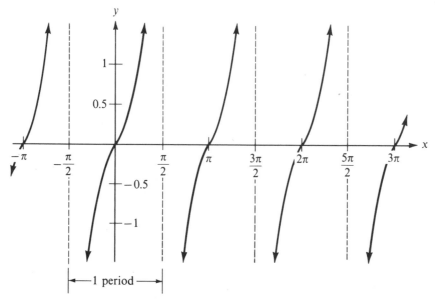

1 period

FIGURE 2.12   Graph of $y = \tan x$.

PROBLEM SET 2.3

A Problems

In Problems 1–6 give the functional value from memory.

1. a. sec 270°   b. cot 3π/2   c. sin 0   d. cos π/6   e. csc 0   f. tan 0°

2. a. csc 30°   b. tan π   c. sec π/2   d. sin π/6   e. cot 180°   f. sin 45°

3. a. tan 90°   b. sec π   c. cot 90°   d. cos 0   e. csc 90°   f. cos π/3

4. a. cot π/6   b. sin 60°   c. cos 45°   d. sec 0°   e. tan π/6   f. csc π

5. a. sin π/2   b. csc 45°   c. tan 60°   d. cot π/4   e. cos π/2   f. sin 270°

6. a. csc 3π/2   b. sin π   c. csc 60°   d. tan 45°   e. cos 90°   f. cot π/6

7. Complete the following table of values for $y = \cos x$:

x = angle	2π/3	3π/4	5π/6	7π/6	5π/4	4π/3	7π/4	11π/6
Quadrant; sign of cos x								
y = cos x								
y approx.								

8. Use the table in Problem 7, along with other values if necessary, to plot $y = \cos x$.

9. Complete a table of values like the one in Problem 7 for $y = \tan x$.

10. Use the table in Problem 9, along with other values if necessary, to plot $y = \tan x$.

11. Draw a quick sketch of $y = \cos x$ from memory by framing the curve. (Note: this is *not* the same thing you did in Problem 8.)

12. Draw a quick sketch of $y = \sin x$ from memory by framing the curve.

13. Draw a quick sketch of $y = \tan x$ from memory by framing the curve.

B Problems

There are other methods for sketching the sine, cosine, and tangent curves besides plotting points or framing the curve. Some of these methods are developed in Problems 14–17.

14. Draw a unit circle, as shown in Figure 2.13. By definition, $\sin \theta = y/r = y/1 = y$. Notice that, as we choose different values for θ, the y-value is the height of the point P above the x-axis. For example, when $\theta = \pi/4$, the point to plot is the intersection of the line $\theta = \pi/4$ and the horizontal line passing through P. As P makes one revolution on the unit circle and the values are plotted in this fashion, one period of the sine curve results. Plot $y = \sin \theta$ in this fashion.

15. Draw a unit circle, as shown in Figure 2.14. By definition, $\cos \theta = x/r = x/1 = x$. Notice the orientation of the axes for the unit circle. Repeat the procedure outlined in Problem 14 to graph $y = \cos \theta$.

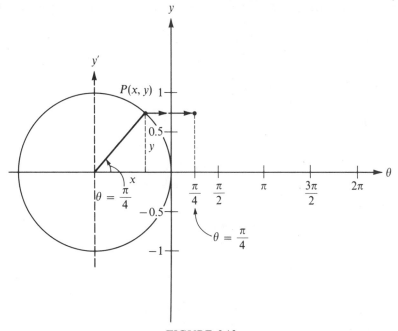

FIGURE 2.13

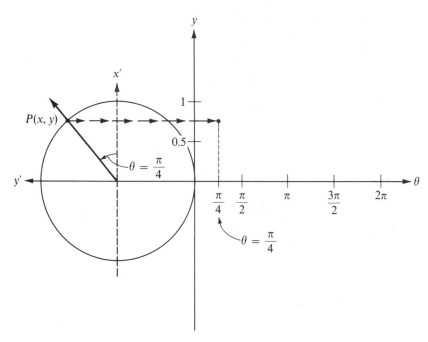

FIGURE 2.14

16. Draw a unit circle with $P(x, y)$ the intersection of the terminal side of an angle θ and the unit circle. Let $|PQ|$ be perpendicular to the x-axis. Draw a line through $S(1, 0)$ perpendicular to the x-axis. Let R be the point of intersection of the line and the terminal side of angle θ. In Problem 14 we showed that $\sin \theta = |PQ|$ and in Problem 15 that $\cos \theta = |OQ|$. Show that $\tan \theta = |RS|$.

17. With the information from Problem 16, graph $y = \tan \theta$ using a technique similar to the one outlined in Problem 14.

Use the technique of plotting points to graph the functions in Problems 18–29.

18. $y = \sec x$	19. $y = \csc x$	20. $y = \cot x$
21. $y = 2 \sin x$	22. $y = \sin 2x$	23. $y = \cos 2x$
24. $y = 2 \sin x + 1$	25. $y = \sin x + \cos x$	26. $y = 2 \tan x$
27. $y = \tan x + \sin x$	28. $y = x + \sin x$	29. $y = \tan 2x$

The graphs of the other trigonometric functions can be found by using the following relationships.

$$\sec \theta = 1/\cos \theta$$
$$\csc \theta = 1/\sin \theta$$
$$\cot \theta = 1/\tan \theta$$

EXAMPLE: Sketch $y = \sec x$.

Solution: Begin by sketching the reciprocal, $y = \cos x$ (dotted curve). Wherever $\cos x = 0$, $\sec x$ is undefined. Draw asymptotes at these places, as shown in Figure 2.15.

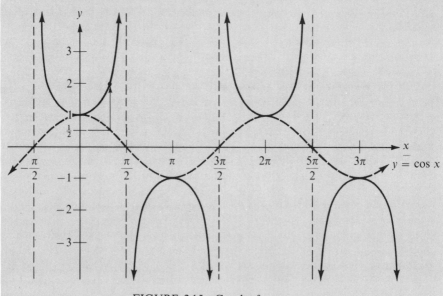

FIGURE 2.15   Graph of $y = \sec x$.

Now plot points by finding the reciprocals of plotted points. For example, when $y = \cos x = \frac{1}{2}$, the reciprocal

$$y = \sec x = \frac{1}{\cos x} = \frac{1}{1/2} = 2.$$

The completed graph is shown.

Graph the curves in Problems 30 and 31 using this technique.

30. $y = \csc x$ 31. $y = \cot x$

Mind Bogglers

32. Plot points to graph $y = 2\sin(3x + \pi/4) + 1$.

33. Plot points to graph $y = (\sin x)/x$.

2.4 GENERAL SINE, COSINE, AND TANGENT
CURVES *(I've Been Framed)*

In Chapter 1 we sketched $y = f(x)$, where f was a linear or quadratic function. We also showed that $y - k = f(x - h)$ is the function that has been translated to the point (h, k). Thus, if we let $f(x) = \sin x$, then $f(x - h) = \sin(x - h)$ is a sine curve shifted h units to the right. (Note: a shift of negative units to the right is a shift to the left.)

EXAMPLE: Graph $y = \sin(x + \pi/2)$.

Solution:

Step 1: Frame the curve as shown in Figure 2.16.
  a. Plot $(h,k) = (-\pi/2, 0)$.
  b. The period of the sine curve is 2π, and it has a high point up one unit and a low point down one unit.

Step 2: Plot the five critical points (two endpoints, the midpoint, and two quarter-points). For the sine curve, plot the endpoint (h, k) and use the frame to plot the other endpoint and the midpoint. For the quarter-points, remember that the sine curve is "up-down"; use the frame to plot the quarter-points as shown in Figure 2.17.

Step 3: Remembering the shape of the sine curve, sketch one period of $y = \sin(x + \pi/2)$ using the frame and the five critical points. If you wish more than one period, just repeat the same pattern.

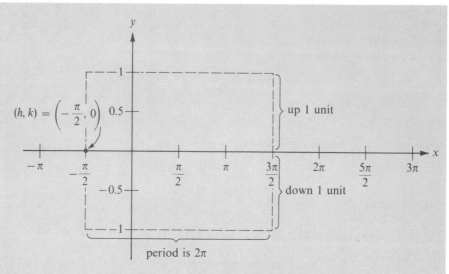

FIGURE 2.16   *Framing the curve*. This step is the same regardless of whether we are graphing a sine, cosine, or tangent curve.

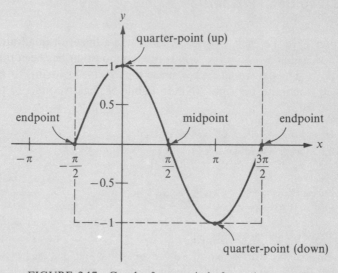

FIGURE 2.17   Graph of one period of $y = \sin(x + \pi/2)$.

Notice from Figure 2.17 that the graph of $y = \sin(x + \pi/2)$ is the same as the graph of $y = \cos x$. Thus

$$\sin(x + \pi/2) = \cos x.$$

EXAMPLE: Graph $y - 2 = \cos(x - \pi/6)$.

Solution:

Step 1: Frame the curve as shown in Figure 2.18; notice that $(h, k) = (\pi/6, 2)$.

Step 2: Plot the five critical points. For the cosine curve the left and right endpoints are at the upper corners of the frame; the midpoint is at the bottom of the frame; the quarter-points are on a line through the middle of the frame.

Step 3: Draw one period of the curve as shown in Figure 2.18.

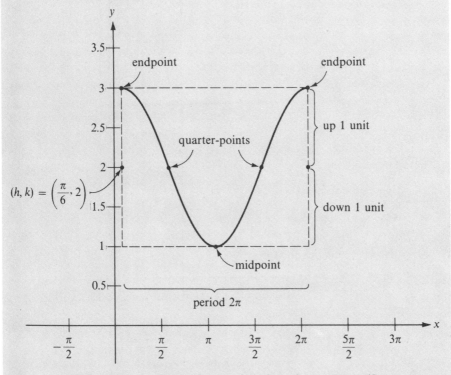

FIGURE 2.18   Graph of one period of $y - 2 = \cos(x - \pi/6)$.

EXAMPLE: Sketch $y + 3 = \tan(x + \pi/3)$.

Solution:

Step 1: Frame the curve as shown in Figure 2.19. Notice that $(h, k) = (-\pi/3, -3)$, and remember that the period of the tangent is π.

Step 2: For the tangent curve, (h, k) is the midpoint of the frame. The endpoints, which are each a distance of $\frac{1}{2}$ the period from the midpoint, determine the location of the asymptotes. The top and the bottom of the frame are one unit from (h, k). Locate the quarter-points at the top and bottom of the frame, as shown in Figure 2.19.

Step 3: Sketch one period of the curve as shown in Figure 2.19. Remember that the tangent curve is not contained within the frame.

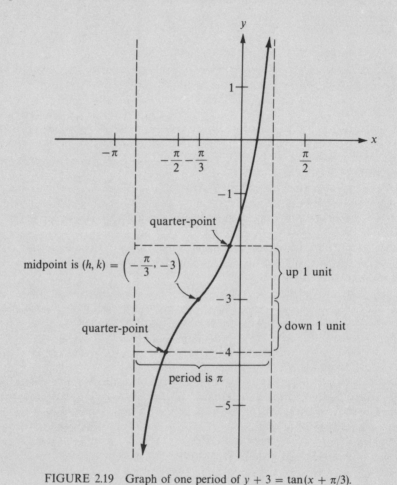

FIGURE 2.19   Graph of one period of $y + 3 = \tan(x + \pi/3)$.

We will discuss two additional changes for the function defined by $y = f(x)$. The first, $y = af(x)$, changes the scale on the y-axis; the second, $y = f(bx)$, changes the scale on the x-axis.

For a function $y = af(x)$ it is clear that the y-value is a times the corresponding value of $f(x)$, which means that $f(x)$ is stretched or shrunk in the y-direction by the multiple of a. For example, if $y = f(x) = \cos x$, then $y = 3f(x) = 3 \cos x$ is the graph of $\cos x$ that has been stretched so that the high point is at 3 units and the low point is at -3 units. In general, given

$$y = af(x),$$

where f represents a trigonometric function, we call $|a|$ the *amplitude* of the function. When $a = 1$, the amplitude is 1 and thus $y = \sin x$, $y = \cos x$, and $y = \tan x$ are all said to have amplitude 1. To graph $y = 3 \cos x$, we frame the cosine curve using an amplitude of 3 rather than 1 (see Figure 2.20).

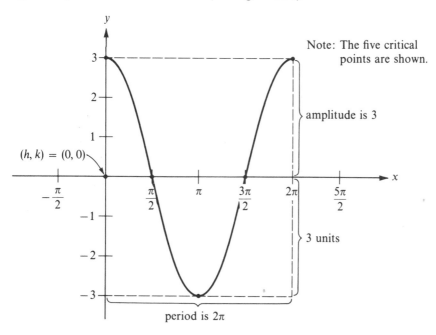

FIGURE 2.20   Graph of one period of $y = 3 \cos x$.

For a function $y = f(bx)$, b affects the scale on the x-axis. Recall that $y = \sin x$ has a period of 2π ($f(x) = \sin x$, so $b = 1$). A function $y = \sin 2x$ ($f(x) = \sin x$ and $f(2x) = \sin 2x$) must complete one period as $2x$ varies from 0 to 2π. This means that one period is completed as x varies from 0 to π. (Remember that for each value of x the result is doubled *before* we find the sine of that number.) In general, the period of $y = \sin bx$ is $2\pi/b$, and the period of $y = \cos bx$ is $2\pi/b$. However, since the period of $y = \tan x$ is π, we can see that $y = \tan bx$ has a period of π/b. Therefore, when framing the curve, we use $2\pi/b$ for the sine and cosine and π/b for the tangent.

EXAMPLE: Graph one period of $y = \sin 2x$.

Solution: The period is $2\pi/2 = \pi$; thus the endpoints of the frame are $(0, 0)$ and $(\pi, 0)$, as shown in Figure 2.21.

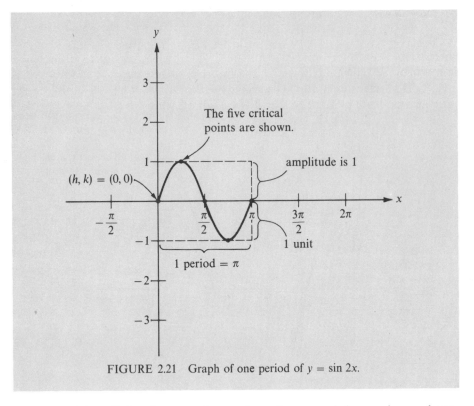

FIGURE 2.21   Graph of one period of $y = \sin 2x$.

Summarizing all the above results, we have the *general* sine, cosine, and tangent curves:

$$y - k = a \sin b(x - h);$$
$$y - k = a \cos b(x - h);$$
$$y - k = a \tan b(x - h).$$

1. The origin has been translated or shifted to the point (h, k).
2. The amplitude is a.
3. The period is $2\pi/b$ for the sine and cosine curves and π/b for the tangent curve.
4. The curves are sketched by translating the origin to the point (h, k) and then framing the curve to complete the graph.

EXAMPLES: Graph the following curves.

1. $y + 1 = 2 \sin \frac{2}{3}(x - \pi/2)$. Notice that $(h, k) = (\pi/2, -1)$ and that the amplitude is 2; the period is $2\pi/(2/3) = 3\pi$. Now plot (h, k) and frame the curve. Then plot the five critical points (two endpoints, the midpoint, and two quarter-points). Finally, after sketching one period, draw the other periods as shown in Figure 2.22.

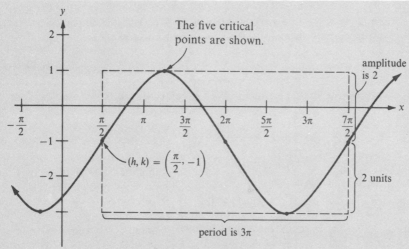

FIGURE 2.22   Graph of $y + 1 = 2 \sin \frac{2}{3}(x - \pi/2)$.

2. $y = 3 \cos(2x + \pi/2) - 2$. Rewrite in standard form to obtain $y + 2 = 3 \cos 2(x + \pi/4)$. Notice that $(h, k) = (-\pi/4, -2)$; the amplitude is 3, and the period is $2\pi/2 = \pi$. Plot (h, k) and frame the curve as shown in Figure 2.23.

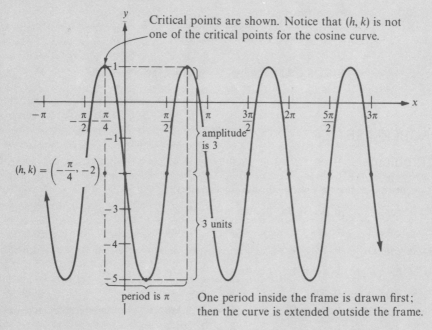

FIGURE 2.23   Graph of $y + 2 = 3 \cos 2(x + \pi/4)$.

3. $y - 2 = 3 \tan \frac{1}{2}(x - \pi/3)$. Notice that $(h, k) = (\pi/3, 2)$; the amplitude is 3; the period is $\pi/(1/2) = 2\pi$. Plot (h, k) and frame the curve as shown in Figure 2.24.

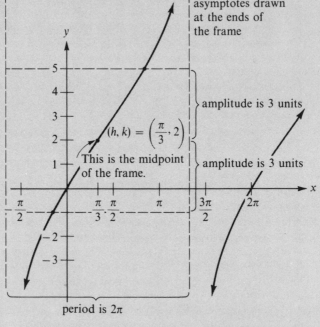

FIGURE 2.24   Graph of $y - 2 = 3 \tan \frac{1}{2}(x - \pi/3)$.

PROBLEM SET 2.4

A Problems

In Problems 1–6 give the functional value from memory.

1. a. $\cos 0$ b. $\sec 3\pi/2$ c. $\tan \pi$ d. $\cot \pi$ e. $\csc 3\pi/2$ f. $\sin 0$

2. a. $\cot \pi/2$ b. $\cos \pi/6$ c. $\sec \pi$ d. $\tan \pi/2$ e. $\sin \pi/6$ f. $\csc \pi$

3. a. $\cot \pi/3$ b. $\csc \pi/6$ c. $\cos \pi/4$ d. $\sin \pi/4$ e. $\tan \pi/3$ f. $\sec \pi/2$

4. a. $\csc \pi/4$ b. $\cot \pi/4$ c. $\sin \pi/2$ d. $\cos \pi/2$ e. $\sec \pi/3$ f. $\tan \pi/4$

5. a. $\cot \pi/6$ b. $\sin \pi$ c. $\csc \pi/3$ d. $\sec \pi/4$ e. $\cos \pi$ f. $\tan \pi/6$

6. a. $\sin 3\pi/2$ b. $\csc \pi/4$ c. $\sec \pi/6$ d. $\tan 2\pi$ e. $\tan 0$ f. $\cos 3\pi/2$

Graph one period of the functions given in Problems 7–20.

7. $y = \sin(x + \pi)$ 8. $y = \cos(x + \pi/2)$

9. $y = \cos(x - 3\pi/2)$　　　　　　　　10. $y = \sin(x - \pi/3)$

11. $y = 3 \sin x$　　　　　　　　　　　　12. $y = 2 \cos x$

13. $y = \sin 3x$　　　　　　　　　　　　　14. $y = \cos 2x$

15. $y = \tan(x - 3\pi/2)$　　　　　　　　16. $y = \tan(x + \pi/6)$

17. $y = \frac{1}{2} \sin x$　　　　　　　　　　18. $y = \frac{1}{3} \tan x$

19. $y = 4 \tan x$　　　　　　　　　　　　20. $y = 5 \sin x$

B Problems

Graph one period of the functions given in Problems 21–30.

21. $y - 2 = \sin(x - \pi/2)$　　　　　　22. $y + 1 = \cos(x + \pi/3)$

23. $y - 3 = \tan(x + \pi/6)$　　　　　　24. $y - \frac{1}{2} = \frac{1}{2} \cos x$

25. $y - 1 = 2 \cos(x - \pi/4)$　　　　　26. $y - 1 = \cos 2(x - \pi/4)$

27. $y + 2 = 3 \sin(x + \pi/6)$　　　　　28. $y + 2 = \sin 3(x + \pi/6)$

29. $y + 2 = \tan(x - \pi/4)$　　　　　　30. $y = 1 + \tan 2(x - \pi/4)$

Graph the curves given in Problems 31–40.

31. $y = \sin(4x + \pi)$　　　　　　　　　32. $y = \sin(3x + \pi)$

33. $y = \tan(2x - \pi/2)$　　　　　　　34. $y = \tan(x/2 + \pi/3)$

35. $y = \frac{1}{2} \cos(x + \pi/6)$　　　　　　36. $y = \cos(\frac{1}{2}x + \pi/12)$

37. $y = 3 \cos(3x + 2\pi) - 2$　　　　　38. $y = 4 \sin(\frac{1}{2}x + 2)$

39. $y = \sqrt{2} \cos(x - \sqrt{2}) - 1$　　　40. $y = \sqrt{3} \sin(\frac{1}{3}x - \sqrt{\frac{1}{3}})$

So far we have limited ourselves to a > 0. If a < 0, the curve is reflected through the x-axis.
Graph the curves in Problems 41–46.

41. $y = -\sin x$　　　　　　　　　　　　42. $y = -\cos x$

43. $y = -\tan x$　　　　　　　　　　　　44. $y = -2 \sin 3x$

45. $y = -3 \cos 2x$　　　　　　　　　　46. $y = -2 \tan(x - \pi/3)$

In Section 1.4 (Problems 36–38) we graphed parabolas by considering the sum of two functions. The same procedure can be used to graph the sum of two trigonometric functions. For example, if

$$y = 3 \sin 2x + 2 \cos 3x,$$

we can consider

$$y = f_1(x) + f_2(x),$$

where $f_1(x) = 3 \sin 2x$ and $f_2(x) = 2 \cos 3x$. We first graph f_1 and f_2 on the same axes, as shown in Figure 2.25. Next, we select a particular x-value—say x_1—and graphically find $y_1 = f_1(x_1)$ and $y_2 = f_2(x_1)$. Then we plot $y = y_1 + y_2$. Do this for several points, as shown, and draw a smooth curve to represent $y = f_1(x) + f_2(x)$.

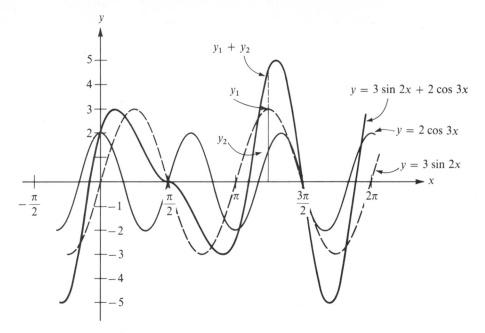

FIGURE 2.25   Graph of $y = 3 \sin 2x + 2 \cos 3x$.

Graph the curves given in Problems 47–52.

47. $y = \sin x + \cos x$ 48. $y = \sin 2x + \cos x$

49. $y = \sin 2x + 2 \sin x$ 50. $y = 2 \cos x + \cos 2x$

51. $y = 3 \sin x + \cos 3x$ 52. $y = 2 \cos x + \sin 2x$

Mind Boggler

53. *What is your biorhythm?* Some people believe that one's entire life can be charted into cycles of good days and bad days. These cycles begin at the moment of birth and continue throughout life. There are three distinct biorhythm charts:

   a. *physical*—a sine curve with a period of 23 days; it shows a cycle of strength, endurance, and energy.

   b. *sensitivity*—a sine curve with a period of 28 days; it shows a cycle of feelings, intuition, moodiness, and creativity.

   c. *mental*—a sine curve with a period of 33 days; it shows a cycle of intelligence, memory, reasoning, power, and reaction.

   On January 25, 1975, an airplane crashed into the radio tower at the American University. The airline's subsequent investigation of the accident included plotting of the pilot's and copilot's biorhythms. These charts are shown in Figure 2.26. Parts above the x-axis are considered positive phases, and parts below are negative phases. The places where the curves cross the x-axis are "critical days." Notice that both the pilot and the copilot were having a "critical day" when the crash occurred and that all but

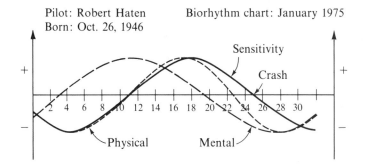

Pilot: Robert Haten   Biorhythm chart: January 1975
Born: Oct. 26, 1946

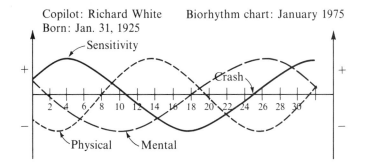

Copilot: Richard White   Biorhythm chart: January 1975
Born: Jan. 31, 1925

FIGURE 2.26   Biorhythm charts.

one of the cycles were in a negative phase, signaling an especially accident-prone time.

a. Write equations for the pilot's and copilot's physical, sensitivity, and mental cycles assuming that each cycle has an amplitude of 1.

b. Calculate your own biorhythm chart. You will first need to find the number of days since your birth. Don't forget that leap years have 366 days. Find the last critical day for each cycle by dividing the number of days since your birth by 23, 28, and 33, respectively.

2.5 INVERSE TRIGONOMETRIC FUNCTIONS (An Arco Sign Is Not Necessarily a Sign for Gasoline)

In Section 1.5 we introduced the notion of the inverse of a function. The ideas presented there apply to trigonometric, as well as algebraic, functions. Recall that, if a function f is given, then its inverse is the set of ordered pairs with the x and y values interchanged. Thus, if

$$y = \sin x$$

is a given function, then its *inverse* is

$$x = \sin y.$$

It is also customary to solve the resulting equation for y. To do this, we define

$$y = \sin^{-1} x \text{ to mean } x = \sin y.$$

That is, $y = $ *the angle whose sine is* x, which is sometimes also written $y = \arcsin x$. It is important to remember that $y = \sin^{-1} x$ is simply a notational change for $x = \sin y$.

In each of the problem sets of this chapter there have been problems in which you were to find the sine of some number—say $\sin 30°$. You have (hopefully) memorized this answer and know that $\sin 30° = \frac{1}{2}$. Now we can reverse the procedure and ask "For which angle θ does $\sin \theta = \frac{1}{2}$?" Using the above notation, we look for $\sin^{-1}\left(\frac{1}{2}\right)$. We see that $\sin^{-1}\left(\frac{1}{2}\right) = 30°$. But is this answer unique? Since $\sin 150° = \frac{1}{2}$, we see that $\sin^{-1}\left(\frac{1}{2}\right) = 150°$. In fact, there are many values of θ for which $\sin \theta = \frac{1}{2}$ (see Figure 2.27). Thus the inverse of $\sin x$—namely $\sin^{-1} x$—is not a function, as we can see by examining the graphs of $y = \sin x$ and $y = \sin^{-1} x$ in Figure 2.28. However, we can define a function $y = \text{Sin } x$ that is identical to $y = \sin x$ except that its domain is limited so that its inverse is a function. From Figure 2.28 we see that, if $y = \text{Sin } x$, where $-\pi/2 \le x \le \pi/2$, then the inverse,

$$y = \text{Sin}^{-1} x, \text{ where } -\pi/2 \le y \le \pi/2,$$

is a function. Thus, $\text{Sin}^{-1}\left(\frac{1}{2}\right) = \pi/6$. The way we have restricted the function is somewhat arbitrary at this point, and other choices might have been made. However, this method is the commonly accepted one in mathematics and is used to simplify later work in this course and in calculus. Remember that our goal was simply to restrict the original function in such a way that its inverse would also be a function that includes all values of the range.

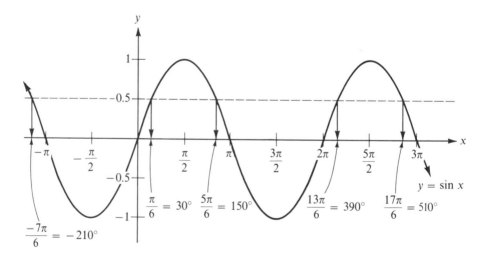

FIGURE 2.27   Graph of $y = \sin x$.

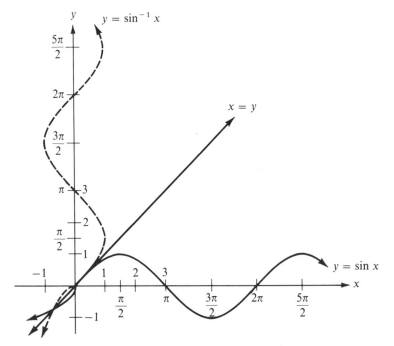

FIGURE 2.28   Graphs of $y = \sin x$ and $y = \sin^{-1} x$.

EXAMPLES: Find the following values.

1. $\mathrm{Sin}^{-1}\left(\frac{1}{2}\sqrt{3}\right) = \theta$.

    Solution: Find the angle θ whose sine is $\frac{1}{2}\sqrt{3}$. Also, since the arcsine is capitalized we want $-\pi/2 \le \theta \le \pi/2$. From the memorized table of values we find $\sin \pi/3 = \frac{1}{2}\sqrt{3}$, and since $\pi/3$ is between $-\pi/2$ and $\pi/2$ we have $\mathrm{Sin}^{-1}\left(\frac{1}{2}\sqrt{3}\right) = \pi/3$.

2. $\mathrm{Sin}^{-1}\left(-\frac{1}{2}\sqrt{3}\right) = \theta$.

    Solution: Find θ where $-\pi/2 \le \theta \le \pi/2$ so that $\sin \theta = -\frac{1}{2}\sqrt{3}$. The sine is negative in both the third and fourth quadrants, but we choose the fourth-quadrant value since $-\pi/2 \le \theta \le \pi/2$. Thus $\mathrm{Sin}^{-1}\left(-\frac{1}{2}\sqrt{3}\right) = -\pi/3$.

3. $\sin^{-1}\left(\frac{1}{2}\sqrt{3}\right) = \theta$.

    Solution: Find θ so that $\sin \theta = \frac{1}{2}\sqrt{3}$. There are many values satisfying this relationship. The sine is positive in the first and second quadrants, so we seek angles in those quadrants whose reference angles are $\pi/3$. Thus $\sin^{-1}\left(\frac{1}{2}\sqrt{3}\right) = \pi/3$ and $\sin^{-1}\left(\frac{1}{2}\sqrt{3}\right) = 2\pi/3$. Also, since the period of the sine is 2π, we also have

    $$\sin^{-1}\left(\tfrac{1}{2}\sqrt{3}\right) = \begin{cases} \pi/3 + 2n\pi \\ 2\pi/3 + 2n\pi \end{cases} \quad \text{where } n \text{ is an integer.}$$

The other trigonometric functions are handled similarly.

Given Function	Inverse	Other Notations for Inverse	
$y = \cos x$	$x = \cos y$	$y = \cos^{-1} x$	$y = \arccos x$
$y = \sin x$	$x = \sin y$	$y = \sin^{-1} x$	$y = \arcsin x$
$y = \tan x$	$x = \tan y$	$y = \tan^{-1} x$	$y = \arctan x$
$y = \sec x$	$x = \sec y$	$y = \sec^{-1} x$	$y = \operatorname{arcsec} x$
$y = \csc x$	$x = \csc y$	$y = \csc^{-1} x$	$y = \operatorname{arccsc} x$
$y = \cot x$	$x = \cot y$	$y = \cot^{-1} x$	$y = \operatorname{arccot} x$

These are the same.

Because the inverse secant and inverse cosecant are rarely used, we will limit our study to the other four inverse trigonometric functions. Consider the graphs in Figure 2.29. We need to restrict each trigonometric function so that the inverse is also a function, but we also want to include all possible values in the range of the original function. For the sine curve we restricted x so that $-\pi/2 \le x \le \pi/2$. Then the inverse is the function

$$y = \operatorname{Sin}^{-1} x, \text{ where } -\pi/2 \le y \le \pi/2.$$

We see that the same restrictions (leaving out the values $-\pi/2$ and $\pi/2$) will apply for the tangent and arctangent curves. However, for the cosine function

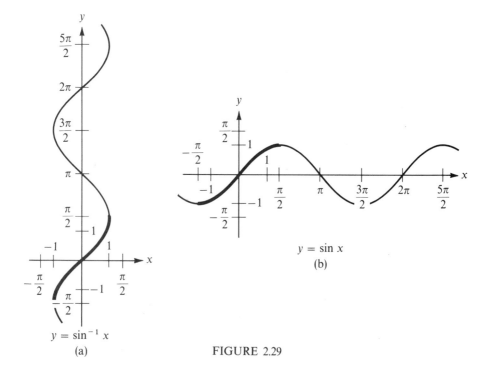

$y = \sin^{-1} x$

(a)

$y = \sin x$

(b)

FIGURE 2.29

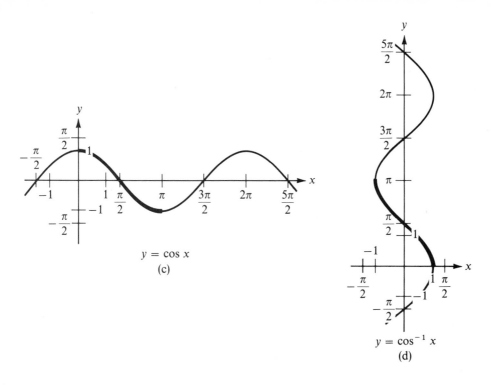

$y = \cos x$
(c)

$y = \cos^{-1} x$
(d)

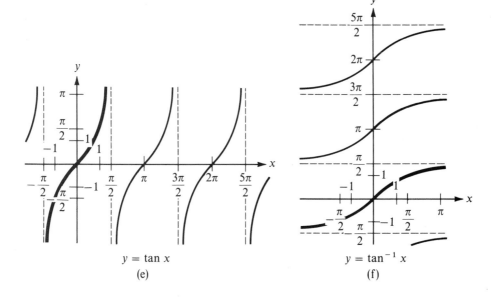

$y = \tan x$
(e)

$y = \tan^{-1} x$
(f)

FIGURE 2.29   (*continued*)

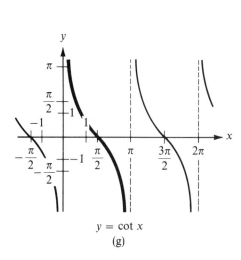

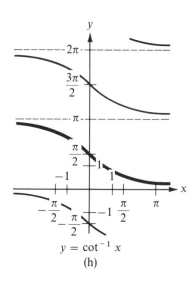

$$y = \cot x$$
(g)

$$y = \cot^{-1} x$$
(h)

FIGURE 2.29   (*continued*)

we see that, by restricting x to the same interval, we obtain only positive values for $y = \cos x$. Thus, to include the entire range of the cosine curve, we must restrict x so that $0 \le x \le \pi$. Then the inverse is the function

$$y = \text{Cos}^{-1} x, \text{ where } 0 \le y \le \pi.$$

We restrict the cotangent function in the same way, and the results are summarized in the definition below. Notice that the inverse functions are indicated in Figure 2.29 by the darker parts of the curves.

Definition:

Inverse Function		Domain	Range
$y = \text{Arccos } x$ or $y = \text{Cos}^{-1} x$		$-1 \le x \le 1$	$0 \le y \le \pi$
$y = \text{Arcsin } x$ or $y = \text{Sin}^{-1} x$		$-1 \le x \le 1$	$-\pi/2 \le y \le \pi/2$
$y = \text{Arctan } x$ or $y = \text{Tan}^{-1} x$		all reals	$-\pi/2 < y < \pi/2$
$y = \text{Arccot } x$ or $y = \text{Cot}^{-1} x$		all reals	$0 < y < \pi$

EXAMPLES: Find the given angle.

1. Arctan $1 = \theta$.

    Solution: We are looking for an angle θ whose tangent is 1. Since this is an exact value, we know that $\theta = \pi/4$ or $45°$.

2. Arccot$(-\sqrt{3}) = \theta$.

Solution: Find θ so that cot $\theta = -\sqrt{3}$; the reference angle is 30°, and the cotangent is negative in Quadrants II and IV. Since Arccotangent is defined in Quadrant II but not in Quadrant IV, we see that Arccot$(-\sqrt{3}) = 150°$.

3. Arcsin$(-0.4695) = \theta$.

Solution: Find θ so that sin $\theta = -0.4695$; from Table III in the back of the book we find that the reference angle is 28°. Since the Arcsine is defined for values between $-90°$ and $90°$, we see that Arcsin$(-0.4695) = -28°$.

Alternate Solution by Calculator: Check the algorithm used in the operation manual of your own calculator.

| enter value | | inv | | trig function |

The | inv | and | trig function | keys pushed in succession give the inverse trig functions. For this example, push the following keys:

| · | | 4 | | 6 | | 9 | | 5 | | +/− | | inv | | sin |

The angle is now displayed: -28.00184535, which is the decimal representation of the angle in degrees.

4. Arccot $\theta = -2.747$.

Solution: We can use Table III to find the reference angle 20°. Since $0 < \cot^{-1} x < \pi$, we must place this angle in Quad II, so $\theta = 160°$. Since calculators don't have a cotangent function, we note that

if cot $y = x$, then tan $y = 1/x$.

Thus,

$$y = \cot^{-1} x \text{ and } y = \tan^{-1}(1/x),$$

$$\cot^{-1} x = \tan^{-1}(1/x).$$

This tells us that to find the inverse cotangent on a calculator we must first take the reciprocal of the given value and then complete the problem. Another difficulty is the negative value since the range for the tangent and cotangent is not the same. Proceed as follows if using a calculator:

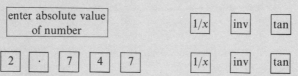

The result 20.00320032 gives the reference angle and thus, the result in Quad II is 160° (159.9967997°).

5.   Arccos(cos $\pi/12$).

Solution: Let $f(x) = \cos x$. Then $f^{-1}(x) = \arccos x$. From Section 1.5, $f^{-1}[f(x)]$ $= f[f^{-1}(x)] = x$. Thus arccos(cos $\pi/12$) $= \pi/12$.

6.   tan(Arctan 2.463) $= 2.463$.

PROBLEM SET 2.5

A Problems

In Problems 1–7 obtain the given angle from memory.

1.   a. Arcsin 0   b. $\text{Tan}^{-1} \sqrt{3}/3$ c. Arccot $\sqrt{3}$ d. Arccos 1

2.   a. $\text{Cos}^{-1} \sqrt{3}/2$ b. Arcsin $\frac{1}{2}$ c. $\text{Tan}^{-1} 1$ d. $\text{Sin}^{-1} 1$

3.   a. Arctan $\sqrt{3}$ b. $\text{Cos}^{-1} \sqrt{2}/2$ c. Arcsin $\frac{1}{2}\sqrt{2}$ d. Arcot 1

4.   a. Arcsin(-1) b. $\text{Cot}^{-1}(-1)$ c. Arcsin$(-\sqrt{3}/2)$ d. $\text{Cos}^{-1}(-1)$

5.   a. $\text{Cot}^{-1}(-\sqrt{3})$ b. Arctan(-1) c. $\text{Sin}^{-1}(-\frac{1}{2}\sqrt{2})$ d. $\text{Cos}^{-1}(-\frac{1}{2})$

6.   a. Arccos$(-\sqrt{2}/2)$ b. $\text{Cot}^{-1}(-\sqrt{3}/3)$ c. $\text{Sin}^{-1}(-\frac{1}{2})$ d. Arctan$(-\sqrt{3}/3)$

7.   a. $\text{Tan}^{-1} 0$ b. Arccot $\sqrt{3}/3$ c. Arccos $\frac{1}{2}$ d. $\text{Sin}^{-1} \sqrt{3}/2$

Use Table III in the back of the book or a calculator to find the values (in angles) given in Problems 8–22.

8.   Arcsin 0.2079   9. $\text{Cos}^{-1} 0.8387$ 10.   Arctan 1.171

11. $\text{Cot}^{-1} 0.0875$ 12. $\text{Tan}^{-1}(-3.732)$ 13.   Arccos 0.9455

14. $\text{Sin}^{-1}(-0.4695)$ 15.   Arccot 0.7265   16.   Arctan 1.036

17. $\text{Cot}^{-1}(-0.3249)$ 18. $\text{Sin}^{-1} 0.3584$ 19. $\text{Cos}^{-1} 0.3584$

20. $\text{Tan}^{-1} 2.050$ 21.   Arcsin(-0.9135) 22.   Arccot(-1.235)

Simplify the expressions in Problems 23–30.

23.   cot(Arccot 1)   24.   Arccos(cos $\pi/6$)

25.   sin(Arcsin 1/3)   26. $\text{Tan}^{-1}(\tan \pi/15)$

27.   Arcsin(sin $2\pi/15$)   28.   cos(Arccos 2/3)

29.   tan(Arctan 0.4163)   30.   Arccot(cot 35°)

B Problems

Find the values given in Problems 31–39.

31. $\sin^{-1} \frac{1}{2}$ 32.   arccos $\frac{1}{2}$ 33. $\tan^{-1} \sqrt{3}$

34. $\cot^{-1}(-\sqrt{3})$ 35. $\cos^{-1}(-\sqrt{3}/2)$ 36.   arcsin$(-\sqrt{2}/2)$

37.   arcsin 0.3907   38.   arccos 0.2924   39.   arctan 1.376

In Problems 40–45 graph the given pair of curves on the same axes.

40. $y = \sin x$; $x = \sin y$ 41. $y = \cos x$; $x = \cos y$ 42. $y = \tan x$; $x = \tan y$

43. $y = \cot x$; $x = \cot y$ 44. $y = \sec x$; $x = \sec y$ 45. $y = \csc x$; $x = \csc y$

Mind Boggler

46. The path of the cartoon character seems to be rather regular or periodic and therefore might be written as a trigonometric function. Suppose each hop is a distance of π ft. Devise an equation that will describe the path of the hops shown.

2.6 SUMMARY AND REVIEW

TERMS

Amplitude (2.4)
Arccosecant (2.5)
Arccosine (2.5)
Arccotangent (2.5)
Arcsecant (2.5)
Arcsine (2.5)
Arctangent (2.5)
Cosecant (2.2)
Cosine (2.1)

Cotangent (2.2)
Frame (2.3, 2.4)
Inverse (2.5)
Period (2.1)
Reduction principle (2.2)
Reference angle (2.2)
Secant (2.2)
Sine (2.1)
Tangent (2.2)

CONCEPTS

(2.2) Let θ be an angle in standard position with a point $P(x, y)$ on the terminal side a distance of $r = \sqrt{x^2 + y^2}$ from the origin ($r \neq 0$). Then we define the **six trigonometric functions** as follows:

$$\cos \theta = \frac{x}{r} \qquad \tan \theta = \frac{y}{x}, \ x \neq 0 \qquad \csc \theta = \frac{r}{y}, \ y \neq 0$$

$$\sin\theta = \frac{y}{r} \qquad \sec\theta = \frac{r}{x},\; x \neq 0 \qquad \cot\theta = \frac{x}{y},\; y \neq 0$$

(2.1) $\cos(\theta \pm 2n\pi) = \cos\theta$, $\sin(\theta \pm 2n\pi) = \sin\theta$, and $\tan(\theta + n\pi) = \tan\theta$. That is, the cosine and sine have period 2π and the tangent has period π.

(2.2) Exact values

function \ angle θ	0	$\dfrac{\pi}{6}$	$\dfrac{\pi}{4}$	$\dfrac{\pi}{3}$	$\dfrac{\pi}{2}$	π	$\dfrac{3\pi}{2}$
$\cos\theta$	1	$\dfrac{\sqrt{3}}{2}$	$\dfrac{\sqrt{2}}{2}$	$\dfrac{1}{2}$	0	-1	0
$\sin\theta$	0	$\dfrac{1}{2}$	$\dfrac{\sqrt{2}}{2}$	$\dfrac{\sqrt{3}}{2}$	1	0	-1
$\tan\theta$	0	$\dfrac{\sqrt{3}}{3}$	1	$\sqrt{3}$	undef.	0	undef.
$\sec\theta$	1	$\dfrac{2}{\sqrt{3}}$	$\dfrac{2}{\sqrt{2}}$	2	undef.	-1	undef.
$\csc\theta$	undef.	2	$\dfrac{2}{\sqrt{2}}$	$\dfrac{2}{\sqrt{3}}$	1	undef.	-1
$\cot\theta$	undef.	$\dfrac{3}{\sqrt{3}}$	1	$\dfrac{\sqrt{3}}{3}$	0	undef.	0

(2.2) $-1 \le \cos\theta \le 1$, $-1 \le \sin\theta \le 1$, $\cos^2\theta + \sin^2\theta = 1$, and $\tan\theta = \sin\theta/\cos\theta$.

(2.2) $\sec\theta = 1/\cos\theta$, $\csc\theta = 1/\sin\theta$, and $\cot\theta = 1/\tan\theta$.

(2.2) **Reduction principle:** If f represents a trigonometric function, then $f(\theta) = \pm f(\theta')$, where θ' is the reference angle for θ and the sign depends on the quadrant and the function:

	y	
Sine positive		All positive
Tangent positive		Cosine positive

(2.3) Framing the cosine, sine, and tangent curves.

(2.4) Graphing the general cosine, sine, and tangent curves:

$$y - k = a\cos b(x - h); \qquad y - k = a\sin b(x - h); \qquad y - k = a\tan b(x - h).$$

We translate the origin to the point (h, k); the amplitude is a; the period is $2\pi/b$ for the cosine and sine curves and π/b for the tangent curve. Next, we frame the curve. Then we plot the critical values and sketch the curve.

(2.5) The **inverse functions** are defined as follows.

Inverse Function	*Domain*	*Range*
$y = \text{Arccos } x$ or $y = \text{Cos}^{-1} x$	$-1 \le x \le 1$	$0 \le y \le \pi$
$y = \text{Arcsin } x$ or $y = \text{Sin}^{-1} x$	$-1 \le x \le 1$	$-\dfrac{\pi}{2} \le y \le \dfrac{\pi}{2}$
$y = \text{Arctan } x$ or $y = \text{Tan}^{-1} x$	all reals	$-\dfrac{\pi}{2} < y < \dfrac{\pi}{2}$
$y = \text{Arccot } x$ or $y = \text{Cot}^{-1} x$	all reals	$0 < y < \pi$

PROBLEM SET 2.6 *(Chapter 2 Test)*

1. (2.2) Give the functional value from memory.
    a. $\cos \pi/3$ b. $\sin \pi/6$ c. $\tan \pi/4$ d. $\sec 0$
    e. $\csc \pi$ f. $\cot \pi/3$ g. $\sin \pi/4$ h. $\tan \pi/3$

2. (2.5) Give the angle from memory.
    a. $\text{Arcsin } 1/2$ b. $\text{Arccos } \sqrt{3}/2$ c. $\text{Tan}^{-1}(-\sqrt{3})$ d. $\text{Arccot } 1$
    e. $\text{Cos}^{-1} 0$ f. $\text{Sin}^{-1}(-\sqrt{2}/2)$ g. $\text{Cot}^{-1}(\tfrac{1}{3}\sqrt{3})$ h. $\text{Arctan } 0$

3. (2.2) Give the functional value from memory, and write the answer in simplest form.
    a. $\sin(\pi/2 - \pi/3)$ b. $\sin \pi/2 \cos \pi/3 - \cos \pi/2 \sin \pi/3$ c. $\sin^2 \pi/6 + \cos^2 \pi/6$

4. (2.2) Give the values of the six trigonometric functions for the following angles.
    a. $-150°$ b. $5\pi/3$ c. passing through $(-4, 1)$

5. (2.2, 2.5) Give the values of the following functions by using Table III or IV or a calculator.
    a. $\sec 23.4°$ b. $\cot 2.5$ c. $\csc 43.28°$
    d. $\text{Arcsin } 0.3140$ e. $\text{Arccos}(-0.6494)$ f. $\text{Arctan } 3.271$

6. (2.3, 2.5) From memory draw a quick sketch of the following curves.
    a. $y = \cos x$ b. $y = \sin x$ c. $y = \tan x$
    d. $y = \arccos x$ e. $y = \arcsin x$ f. $y = \arctan x$

7. (2.4) Graph the following curves.
    a. $y - 2 = \sin(x - \pi/6)$ b. $y = \cos(x + \pi/4)$ c. $y = \tan(x - \pi/3) - 2$

8. (2.4) Graph the following curves.
    a. $y = 2 \cos \tfrac{2}{3}x$ b. $y = \tfrac{1}{3} \sin 2x$ c. $y = \tfrac{1}{2} \tan \tfrac{1}{2}x$

9. (2.4) Graph the following curves.
    a. $y = 2 \cos(2x + \pi/3)$ b. $y = 2 \sin 2x + \cos x$

10. (2.5) Find the indicated values.
    a. $\sin^{-1}(-\tfrac{1}{2}\sqrt{3})$ b. $\cos(\text{Cos}^{-1} \tfrac{4}{5})$ c. $\arctan 2.475$

In this chapter we will use the definitions of the six trigonometric functions to develop *eight fundamental identities,* which can be employed to change the form of a trigonometric expression. These identities will then be used to derive a variety of other useful identities. Historically, trigonometry was developed as a manipulative subject whose primary purpose was to allow people to solve certain relationships concerning triangles. Although this goal is still important (see Chapter 5), computers and calculators have minimized the manipulative aspects of trigonometry. Today much more stress is placed on the character of the functions themselves; it is important to be able to change the form of a trigonometric expression in many different ways. Anyone who is planning further study in mathematics, engineering, or physics will need to become thoroughly familiar with the material of this chapter.

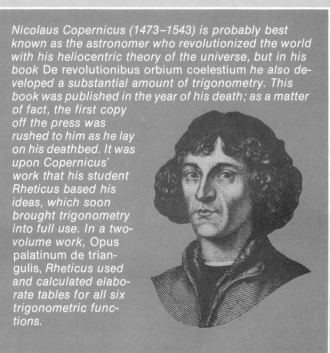

Nicolaus Copernicus (1473–1543) is probably best known as the astronomer who revolutionized the world with his heliocentric theory of the universe, but in his book De revolutionibus orbium coelestium *he also developed a substantial amount of trigonometry. This book was published in the year of his death; as a matter of fact, the first copy off the press was rushed to him as he lay on his deathbed. It was upon Copernicus' work that his student Rheticus based his ideas, which soon brought trigonometry into full use. In a two-volume work,* Opus palatinum de triangulis, *Rheticus used and calculated elaborate tables for all six trigonometric functions.*

3

TRIGONOMETRIC IDENTITIES

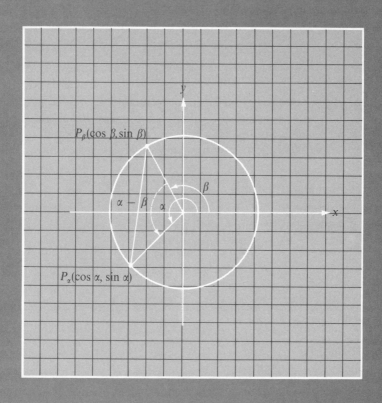

3.1 EQUATIONS AND IDENTITIES *(The Main Problem Is to Establish Your Identity)*

In mathematics we consider several types of equations, which are summarized as follows.

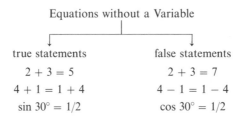

Equations without a Variable

true statements	false statements
$2 + 3 = 5$	$2 + 3 = 7$
$4 + 1 = 1 + 4$	$4 - 1 = 1 - 4$
$\sin 30° = 1/2$	$\cos 30° = 1/2$

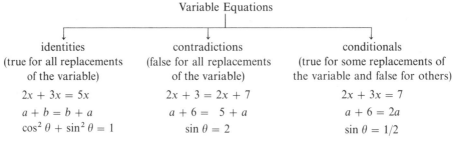

Variable Equations

identities (true for all replacements of the variable)	contradictions (false for all replacements of the variable)	conditionals (true for some replacements of the variable and false for others)
$2x + 3x = 5x$	$2x + 3 = 2x + 7$	$2x + 3x = 7$
$a + b = b + a$	$a + 6 = 5 + a$	$a + 6 = 2a$
$\cos^2 \theta + \sin^2 \theta = 1$	$\sin \theta = 2$	$\sin \theta = 1/2$

In arithmetic you focused your attention primarily on true equations (hopefully you didn't write too many false equations), whereas in algebra you concentrated on conditional equations. However, in order to solve equations, you first had to learn some identities (sometimes called laws or properties), such as the following.

Commutativity: $a + b = b + a$; $ab = ba$

Associativity: $a + (b + c) = (a + b) + c$; $a(bc) = (ab)c$

Distributivity: $a(b + c) = ab + ac$

In trigonometry we will also study identities, as well as conditional equations. Recall that:

An *equation* consists of two members connected by an equals sign ($=$). We call these members the left-hand side (L) and the right-hand side (R), so that the equation has the form

L = R.

If the equation contains at least one variable, it is called a *variable equation*.

A *conditional equation* is a variable equation that is true for some (at least one), but not all, replacements of the variable for which the members of the equation are defined.

An *identity* is a variable equation that is true for all replacements of the variable for which the members of the equation are defined.

All of our work with trigonometric identities is ultimately based on eight basic identities, called the *fundamental identities*. Notice that these identities are classified into three categories and are numbered for later reference.

FUNDAMENTAL IDENTITIES

Reciprocal Identities

$$\sec \theta = \frac{1}{\cos \theta}, \quad \cos \theta \neq 0 \qquad (1)$$

$$\csc \theta = \frac{1}{\sin \theta}, \quad \sin \theta \neq 0 \qquad (2)$$

$$\cot \theta = \frac{1}{\tan \theta}, \quad \tan \theta \neq 0 \qquad (3)$$

Ratio Identities

$$\tan \theta = \frac{\sin \theta}{\cos \theta}, \quad \cos \theta \neq 0 \qquad (4)$$

$$\cot \theta = \frac{\cos \theta}{\sin \theta}, \quad \sin \theta \neq 0 \qquad (5)$$

Pythagorean Identities

$$\sin^2 \theta + \cos^2 \theta = 1 \qquad (6)$$

$$1 + \tan^2 \theta = \sec^2 \theta \qquad (7)$$

$$1 + \cot^2 \theta = \csc^2 \theta \qquad (8)$$

The proofs of these identities follow directly from the definitions of the trigonometric functions. Some proofs were given in the previous chapter, but they are repeated here and in Problem Set 3.1.

PROOFS OF THE FUNDAMENTAL IDENTITIES

Let θ be an angle in standard position with point $P(x, y)$ on the terminal side a distance of r from the origin, with $r \neq 0$.

(1) $\sec \theta = 1/\cos \theta$

By definition, $\cos \theta = x/r$; thus

$$\frac{1}{\cos \theta} = \frac{1}{x/r}$$

$$= 1 \cdot \frac{r}{x} \qquad \text{Division of fractions}$$

$$= \frac{r}{x} \qquad \text{Multiplication of fractions}$$

$$= \sec \theta. \qquad \text{By definition of } \sec \theta$$

(2)   and (3) are proved in precisely the same way and are left as problems.

(4) $\tan \theta = \sin \theta / \cos \theta$

$$\frac{\sin \theta}{\cos \theta} = \frac{y/r}{x/r} \qquad \text{By definition of } \sin \theta \text{ and } \cos \theta$$

$$= \frac{y}{r} \cdot \frac{r}{x} \qquad \text{Division of fractions}$$

$$= \frac{y}{x} \qquad \text{Multiplication and simplification of fractions}$$

$$= \tan \theta \qquad \text{By definition of } \tan \theta$$

(5) $\cot \theta = \cos \theta / \sin \theta$. This identity can be proved by using the definition of the trigonometric functions in the same manner as (4) above. However, we can also use previously proved identities.

$$\cot \theta = \frac{1}{\tan \theta} \qquad \text{By identity (3)}$$

$$= \frac{1}{\sin \theta / \cos \theta} \qquad \text{Identity (4)}$$

$$= 1 \cdot \frac{\cos \theta}{\sin \theta} \qquad \text{Division of fractions}$$

$$= \frac{\cos \theta}{\sin \theta} \qquad \text{Multiplication of fractions}$$

(6) $\sin^2 \theta + \cos^2 \theta = 1$. In proving identities, it is important not to start with the given identity, because we don't know it is true until we prove it true! Notice that in all the above examples we *began* with something we knew was true and *ended* with the identity we were trying to prove. For identities (6), (7), and (8) we begin with the Pythagorean Theorem (which is why these are called the Pythagorean identities).

$$x^2 + y^2 = r^2 \qquad \text{By the Pythagorean Theorem}$$

To prove (6), we divide both sides by r^2; for (7) we divide by x^2 and for (8) by y^2. We will show the details for (6) and leave (7) and (8) as problems.

$$\frac{x^2}{r^2} + \frac{y^2}{r^2} = \frac{r^2}{r^2} \qquad \text{Dividing both sides by } r^2 (r \neq 0)$$

$$\left(\frac{x}{r}\right)^2 + \left(\frac{y}{r}\right)^2 = 1 \qquad \text{Properties of exponents}$$

$$(\cos \theta)^2 + (\sin \theta)^2 = 1 \qquad \text{Definition of } \cos \theta \text{ and } \sin \theta$$

$$\sin^2 \theta + \cos^2 \theta = 1 \qquad \text{Commutative property}$$

These eight identities, as well as their various equivalent forms, should be memorized.

Identity	*Equivalent Forms*

(1) $\sec \theta = \dfrac{1}{\cos \theta}$ $\cos \theta \sec \theta = 1$; $\sec \theta \cos \theta = 1$; $\cos \theta = \dfrac{1}{\sec \theta}$

(2) $\csc \theta = \dfrac{1}{\sin \theta}$ $\sin \theta \csc \theta = 1$; $\csc \theta \sin \theta = 1$; $\sin \theta = \dfrac{1}{\csc \theta}$

(3) $\cot \theta = \dfrac{1}{\tan \theta}$ $\tan \theta \cot \theta = 1$; $\cot \theta \tan \theta = 1$; $\tan \theta = \dfrac{1}{\cot \theta}$

(4) $\tan \theta = \dfrac{\sin \theta}{\cos \theta}$ $\cos \theta \tan \theta = \sin \theta$; $\tan \theta \cos \theta = \sin \theta$; $\cos \theta = \dfrac{\sin \theta}{\tan \theta}$

(5) $\cot \theta = \dfrac{\cos \theta}{\sin \theta}$ $\sin \theta \cot \theta = \cos \theta$; $\cot \theta \sin \theta = \cos \theta$; $\sin \theta = \dfrac{\cos \theta}{\cot \theta}$

(6) $\sin^2 \theta + \cos^2 \theta = 1$ $\cos^2 \theta + \sin^2 \theta = 1$; $\sin^2 \theta = 1 - \cos^2 \theta$; $\cos^2 \theta = 1 - \sin^2 \theta$;
$\sin \theta = \pm\sqrt{1 - \cos^2 \theta}$; $\cos \theta = \pm\sqrt{1 - \sin^2 \theta}$

(7) $1 + \tan^2 \theta = \sec^2 \theta$ $\tan^2 \theta = \sec^2 \theta - 1$; $\tan \theta = \pm\sqrt{\sec^2 \theta - 1}$;
$\sec \theta = \pm\sqrt{1 + \tan^2 \theta}$; $\sec^2 \theta - \tan^2 \theta = 1$

(8) $1 + \cot^2 \theta = \csc^2 \theta$ $\cot^2 \theta = \csc^2 \theta - 1$; $\cot \theta = \pm\sqrt{\csc^2 \theta - 1}$;
$\csc \theta = \pm\sqrt{1 + \cot^2 \theta}$; $\csc^2 \theta - \cot^2 \theta = 1$

EXAMPLE: Write all the trigonometric functions in terms of $\sin \theta$.

Solution: a. $\sin \theta = \sin \theta$

b. $\cos \theta = \pm\sqrt{1 - \sin^2 \theta}$ From (6)

c. $\tan \theta = \dfrac{\sin \theta}{\cos \theta}$ From (4)

$\qquad = \dfrac{\sin \theta}{\pm\sqrt{1 - \sin^2 \theta}}$ From part (b)

d. $\cot \theta = \dfrac{1}{\tan \theta}$ From (3)

$\qquad = \dfrac{\pm\sqrt{1 - \sin^2 \theta}}{\sin \theta}$ From part (c)

e. $\csc \theta = \dfrac{1}{\sin \theta}$ From (2)

f. $\sec \theta = \dfrac{1}{\cos \theta}$ From (1)

$\qquad = \dfrac{1}{\pm\sqrt{1 - \sin^2 \theta}}$ From part (b)

The $\pm$ sign we've been using, as in

$$\cos\theta = \pm\sqrt{1 - \sin^2\theta},$$

means that $\cos\theta$ is positive for some values of θ and negative for other values of θ. The $+$ or the $-$ sign is chosen by determining the proper quadrant, as shown in the following example.

EXAMPLE: Given $\sin\theta = 3/5$ and $\tan\theta < 0$, find the other functions of θ.

Solution: Since the tangent is negative and the sine is positive, we see that the proper quadrant is II. Thus,

$$\cos\theta = -\sqrt{1 - \sin^2\theta} \qquad \text{(since the cosine is negative in Quadrant II)}$$

$$= -\sqrt{1 - (3/5)^2}$$

$$= -\sqrt{1 - 9/25}$$

$$= -\sqrt{16/25}$$

$$= -4/5.$$

Also,

$$\tan\theta = \frac{\sin\theta}{\cos\theta} = \frac{3/5}{-4/5} = -\frac{3}{4}.$$

Using the reciprocal identities, we find $\cot\theta = -4/3$, $\sec\theta = -5/4$, and $\csc\theta = 5/3$.

PROBLEM SET 3.1

A Problems

1. State from memory the eight fundamental identities.

*In Problems 2–9 state the **quadrant(s)** in which θ may lie to make the expression true.*

2. $\sin\theta = \sqrt{1 - \cos^2\theta}$

3. $\sin\theta = -\sqrt{1 - \cos^2\theta}$

4. $\sec\theta = -\sqrt{1 + \tan^2\theta}$

5. $\sec\theta = \sqrt{1 + \tan^2\theta}$

6. $\csc\theta = \sqrt{1 + \cot^2\theta}$; $\tan\theta < 0$

7. $\cos\theta = -\sqrt{1 - \sin^2\theta}$; $\sin\theta > 0$

8. $\tan\theta = \sqrt{\sec^2\theta - 1}$; $\cos\theta < 0$

9. $\cot\theta = \sqrt{1 + \cot^2\theta}$; $\sin\theta > 0$

Write each of the expressions in Problems 10–19 as a single trigonometric function of some angle by using one of the eight fundamental identities.

10. $\dfrac{\sin 50°}{\cos 50°}$

11. $\dfrac{\cos(A + B)}{\sin(A + B)}$

12. $\dfrac{1}{\sec 75°}$

13. $\dfrac{1}{\cot(\pi/15)}$

14. $\tan 42° \cos 42°$

15. $\cot(\pi/8)\sin(\pi/8)$

16. $1 - \cos^2 18°$

17. $-\sqrt{1 - \sin^2 127°}$

18. $\sec^2(\pi/6) - 1$

19. $1 + \tan^2(\pi/5)$

Evaluate the expressions in Problems 20–25 by using one of the eight fundamental identities.

20. $\cos 128° \sec 128°$

21. $\sin^2 \pi/3 + \cos^2 \pi/3$

22. $\sec^2 \pi/6 - \tan^2 \pi/6$

23. $\cot^2 45° - \csc^2 45°$

24. $\tan^2 135° - \sec^2 135°$

25. $\csc 85° \sin 85°$

26. Prove that $\csc \theta = 1/\sin \theta$.

27. Prove that $\cot \theta = 1/\tan \theta$.

28. Prove that $1 + \tan^2 \theta = \sec^2 \theta$.

29. Prove that $1 + \cot^2 \theta = \csc^2 \theta$.

B Problems

In Problems 30–34 write all the trigonometric functions in terms of the given function.

30. $\cos \theta$ 31. $\tan \theta$ 32. $\cot \theta$ 33. $\sec \theta$ 34. $\csc \theta$

In Problems 35–44 find the other functions of θ using the given information.

35. $\cos \theta = 5/13;\ \tan \theta < 0$

36. $\cos \theta = 5/13;\ \tan \theta > 0$

37. $\tan \theta = 5/12;\ \sin \theta > 0$

38. $\tan \theta = 5/12;\ \sin \theta < 0$

39. $\sin \theta = 2/3;\ \sec \theta > 0$

40. $\sin \theta = 2/3;\ \sec \theta < 0$

41. $\sec \theta = \sqrt{34}/5;\ \tan \theta < 0$

42. $\sec \theta = \sqrt{34}/5;\ \tan \theta > 0$

43. $\csc \theta = -\sqrt{10}/3;\ \cos \theta > 0$

44. $\csc \theta = -\sqrt{10}/3;\ \cos \theta < 0$

Simplify the expressions in Problems 45–50 by using the fundamental identities.

45. $\dfrac{1 - \cos^2 \theta}{\sin \theta}$

46. $\dfrac{1 - \sin^2 \theta}{\cos \theta}$

47. $\dfrac{\dfrac{\sin \theta}{\cos \theta} + \dfrac{\cos \theta}{\sin \theta}}{\dfrac{1}{\sin \theta \cos \theta}}$

48. $\sin \theta + \dfrac{\cos^2 \theta}{\sin \theta}$

49. $\dfrac{\cos \theta + \dfrac{\sin^2 \theta}{\cos \theta}}{\sin \theta}$

50. $\dfrac{\dfrac{\cos^4 \theta}{\sin^2 \theta} + \cos^2 \theta}{\dfrac{\cos^2 \theta}{\sin^2 \theta}}$

Reduce the expressions in Problems 51–58 so that they involve only sines and cosines, and then simplify.

51. $\sin \theta + \cot \theta$

52. $\sec \theta + \tan \theta$

53. $\dfrac{\tan \theta + \cot \theta}{\sec \theta \csc \theta}$

54. $\dfrac{\sec \theta + \csc \theta}{\tan \theta \cot \theta}$

55. $\sec^2 \theta + \tan^2 \theta$

56. $\csc^2 \theta + \cot^2 \theta$

57. $(\cot \theta - \sec \theta)(\sin \theta \cos \theta)$

58. $(\tan \theta - \csc \theta)(\cos \theta \sin \theta)$

Mind Bogglers

59. a. Explain why the following "number trick" works: Pick any number. Square the number. Add 25. Add 10 times the original number. Divide by 5 more than the original number. Subtract your original number. Your answer is 5.
   b. What term introduced in this section can be applied to your explanation of part (a)?
   c. Will this trick always work?

60. Consider the following problem: An elephant and a bird wish to play together on a teeter-totter. The bird says that it is an impossible idea, but the elephant assures the little bird that it will work out and that he will prove it, since he has had a little algebra.

He presents the following argument to the little bird:

Let E = the weight of the elephant;

b = the weight of the bird.

Now there must be some weight, w (probably very large), so that

$E = b + w.$

Multiply both sides by $E - b$:

$$E(E - b) = (b + w)(E - b)$$
$$E^2 - Eb = bE + wE - b^2 - wb$$
$$E^2 - Eb - wE = bE - b^2 - wb \quad \text{(subtracting } wE \text{ from both sides)}$$
$$E(E - b - w) = b(E - b - w)$$
$$E = b \quad \text{(dividing both sides by } E - b - w)$$

This last equation says that the weight of the elephant is the same as the weight of the bird. "Now," says the elephant, "since our weights are the same, we'll have no problem on the teeter-totter."

"Wait!" hollers the bird. "Obviously this is false." But where is the error in the reasoning?

3.2 PROVING IDENTITIES (It Takes Two Secs to Calculate the Tan)

In the last section we considered eight fundamental identities, which are used to simplify and change the form of a great variety of trigonometric expressions. Suppose we're given a trigonometric equation such as

$$\tan \theta + \cot \theta = \sec \theta \csc \theta$$

and are asked to show that it is an identity. We must be careful to treat this problem differently than we would if it were an algebraic equation. When asked to prove an identity, we do *not* start with the given expression, since we cannot assume that it is true. What, then, can we do if we cannot begin with the given equation? We can begin with what we know is true and *end* with the given identity. This means that, when we are given an identity to prove such as

$$L = R,$$

there are three ways of proceeding:

1. Reduce the left-hand side to the right-hand side by using algebra and the fundamental identities. Thus

    $$L = L$$
    $$= L_1$$
    $$= L_2$$
    $$\vdots$$
    $$= L_n$$

    so that $L_n = R$ after n reversible steps.

2. Reduce the right-hand side to the left-hand side. Thus

    $$R = R$$
    $$= R_1$$
    $$= R_2$$
    $$\vdots$$
    $$= R_n$$

    so that $R_n = L$.

3. Reduce both sides independently to the same expression.

$L = L$	$R = R$
$= L_1$	$= R_1$
$= L_2$	$= R_2$
$\vdots$	$\vdots$
$= L_m$	$= R_m$

    If these two are identical,
    then $L = R$.

We'll illustrate these techniques with several examples.

EXAMPLE: Prove that $\tan\theta + \cot\theta = \sec\theta\csc\theta$.

Method-I solution: Reduce the left-hand side to the right-hand side.

$$\tan\theta + \cot\theta = \frac{\sin\theta}{\cos\theta} + \frac{\cos\theta}{\sin\theta}$$

$$= \frac{\sin^2\theta}{\cos\theta\sin\theta} + \frac{\cos^2\theta}{\cos\theta\sin\theta}$$

$$= \frac{\sin^2\theta + \cos^2\theta}{\cos\theta\sin\theta}$$

$$= \frac{1}{\cos\theta\sin\theta}$$

$$= \frac{1}{\cos\theta} \cdot \frac{1}{\sin\theta}$$

$$= \sec\theta\csc\theta$$

Note: In proving identities, it is often advantageous to change all the trigonometric functions involved to sines and cosines.

Thus, $\tan\theta + \cot\theta = \sec\theta\csc\theta$.

Note: When we arrive at the given identity, we are finished with the problem.

Method-II solution: Reduce the right-hand side to the left-hand side.

$$\sec\theta\csc\theta = \frac{1}{\cos\theta} \cdot \frac{1}{\sin\theta}$$

$$= \frac{1}{\cos\theta\sin\theta}$$

$$= \frac{\sin^2\theta + \cos^2\theta}{\cos\theta\sin\theta}$$

$$= \frac{\sin^2\theta}{\cos\theta\sin\theta} + \frac{\cos^2\theta}{\cos\theta\sin\theta}$$

$$= \frac{\sin\theta}{\cos\theta} + \frac{\cos\theta}{\sin\theta}$$

$$= \tan\theta + \cot\theta$$

Thus, $\tan\theta + \cot\theta = \sec\theta\csc\theta$.

Method-III solution: Reduce both sides to a form that is the same.

$$\tan\theta + \cot\theta = \frac{\sin\theta}{\cos\theta} + \frac{\cos\theta}{\sin\theta} \qquad \sec\theta\csc\theta = \frac{1}{\cos\theta} \cdot \frac{1}{\sin\theta}$$

$$= \frac{\sin^2\theta + \cos^2\theta}{\cos\theta\sin\theta} \qquad = \frac{1}{\cos\theta\sin\theta}$$

$$= \frac{1}{\cos\theta\sin\theta}$$

These are identical.

Therefore, $\tan\theta + \cos\theta = \sec\theta\csc\theta$.

Of course, we would pick only one of these methods when proving a particular identity. Generally, it is easier to begin with the more complicated side and try to reduce it to the simpler side. If both sides seem equally simple, you might change all the functions to sines and cosines and then simplify.

EXAMPLE: Prove that $1/(1 + \cos \theta) + 1/(1 - \cos \theta) = 2 \csc^2 \theta$.

Solution: We begin with the more complicated side.

$$\frac{1}{1 + \cos \theta} + \frac{1}{1 - \cos \theta} = \frac{(1 - \cos \theta) + (1 + \cos \theta)}{(1 + \cos \theta)(1 - \cos \theta)}$$

$$= \frac{2}{1 - \cos^2 \theta}$$

$$= \frac{2}{\sin^2 \theta}$$

$$= 2 \csc^2 \theta$$

EXAMPLE: Prove that $\tan 4\theta + \cot 4\theta = \sec 4\theta \csc 4\theta$.

Solution: Both sides seem equally simple, so we change to sines and cosines and then simplify.

$$\tan 4\theta + \cot 4\theta = \frac{\sin 4\theta}{\cos 4\theta} + \frac{\cos 4\theta}{\sin 4\theta}$$

$$= \frac{\sin^2 4\theta + \cos^2 4\theta}{\cos 4\theta \sin 4\theta}$$

$$= \frac{1}{\cos 4\theta \sin 4\theta}$$

$$= \frac{1}{\cos 4\theta} \cdot \frac{1}{\sin 4\theta}$$

$$= \sec 4\theta \csc 4\theta$$

EXAMPLE: Prove that $(\sec 2\lambda + \cot 2\lambda)/\sec 2\lambda = 1 + \csc 2\lambda - \sin 2\lambda$.

Solution: We begin with the left-hand side. When we are working with a fraction consisting of a single function as a denominator, it is often helpful to separate the fraction into the sum of several fractions.

$$\frac{\sec 2\lambda + \cot 2\lambda}{\sec 2\lambda} = \frac{\sec 2\lambda}{\sec 2\lambda} + \frac{\cot 2\lambda}{\sec 2\lambda}$$

$$= 1 + \cot 2\lambda \cdot \frac{1}{\sec 2\lambda}$$

$$= 1 + \frac{\cos 2\lambda}{\sin 2\lambda} \cdot \cos 2\lambda$$

$$= 1 + \frac{\cos^2 2\lambda}{\sin 2\lambda}$$

$$= 1 + \frac{1 - \sin^2 2\lambda}{\sin 2\lambda}$$

$$= 1 + \frac{1}{\sin 2\lambda} - \frac{\sin^2 2\lambda}{\sin 2\lambda}$$

$$= 1 + \csc 2\lambda - \sin 2\lambda$$

We will consider some additional simplifications of trigonometric identities in the next section.

PROBLEM SET 3.2

A Problems

Prove the following identities.

1. $\tan \theta = \sin \theta \sec \theta$

2. $\cot \theta = \cos \theta \csc \theta$

3. $\sec \theta = \dfrac{\cos \theta}{1 - \sin^2 \theta}$

4. $\csc \theta = \dfrac{\sin \theta}{1 - \cos^2 \theta}$

5. $\tan 2\theta \csc 2\theta = \sec 2\theta$

6. $\cos 2\theta = \sin 2\theta \cot 2\theta$

7. $\tan \theta \cos \theta = \sin \theta$

8. $\cot \theta \sin \theta = \cos \theta$

9. $\dfrac{\sin \theta \csc \theta}{\tan \theta} = \cot \theta$

10. $\dfrac{\cot \theta \tan \theta}{\sin \theta} = \csc \theta$

11. $\dfrac{\cot t}{\sin^2 t + \cos^2 t} = \dfrac{1}{\tan t}$

12. $\dfrac{\sec^2 t}{1 + \tan^2 t} = 1$

13. $\dfrac{\sec^2 u}{\tan^2 u} = \csc^2 u$

14. $\dfrac{\sec u}{\tan^2 u} = \cot u \csc u$

15. $\sec^2 \gamma + 2 \tan \gamma = \dfrac{\sec^2 \gamma \cos \gamma + 2 \sin \gamma}{\cos \gamma}$

16. $\sec \gamma + \tan \gamma = \dfrac{\csc \gamma + 1}{\csc \gamma \cos \gamma}$

17. $\tan 3\theta \sin 3\theta = \sec 3\theta - \cos 3\theta$

18. $\csc 3\theta - \sin 3\theta = \cot 3\theta \cos 3\theta$

19. $1 + \sin^2 \lambda = 2 - \cos^2 \lambda$

20. $2 - \sin^2 3\lambda = 1 + \cos^2 3\lambda$

21. $\dfrac{\sin \alpha}{\tan \alpha} + \dfrac{\cos \alpha}{\cot \alpha} = \cos \alpha + \sin \alpha$

22. $\dfrac{1}{1 + \cos 2\alpha} + \dfrac{1}{1 - \cos 2\alpha} = 2 \csc^2 2\alpha$

23. $\sec \beta + \cos \beta = \dfrac{2 - \sin^2 \beta}{\cos \beta}$

24. $2 \sin^2 3\beta - 1 = 1 - 2 \cos^2 3\beta$

25. $(\tan 5\beta - 1)(\tan 5\beta + 1) = \sec^2 5\beta - 2$

26. $\tan^2 7\alpha = (\sec 7\alpha - 1)(\sec 7\alpha + 1)$

27. $(\sin \gamma - \cos \gamma)^2 = 1 - 2 \sin \gamma \cos \gamma$

28. $(\sin \gamma + \cos \gamma)(\sin \gamma - \cos \gamma) = 2 \sin^2 \gamma - 1$

29. $\dfrac{\sin \gamma}{\cos \gamma} + \cot \gamma = \sec \gamma \csc \gamma$

30. $\tan 3\gamma + \cot 3\gamma = \sec 3\gamma \csc 3\gamma$

B Problems

31. $\csc \theta + \sin \theta = 2 \csc \theta - \cot \theta \cos \theta$

32. $\csc \theta - \sin \theta = \cos^2 \theta \csc \theta$

33. $(\csc 2\theta - \sin 2\theta)(\csc 2\theta + \sin 2\theta) = \csc^2 2\theta(1 - \sin^4 2\theta)$

34. $\dfrac{\tan 2\theta + \cot 2\theta}{\sec 2\theta} = \csc 2\theta$

35. $\dfrac{\tan 3\theta + \cot 3\theta}{\csc 3\theta} = \sec 3\theta$

36. $\dfrac{\sec \lambda + \tan^2 \lambda}{\sec \lambda} = 1 + \sec \lambda - \cos \lambda$

37. $\dfrac{\sin 2\lambda}{\tan 2\lambda} + \dfrac{\cos 2\lambda}{\cot 2\lambda} = \cos 2\lambda + \sin 2\lambda$

38. $\dfrac{\cot 3\theta - \sin 3\theta}{\sin 3\theta} = \dfrac{\cos^2 3\theta + \cos 3\theta - 1}{\sin^2 3\theta}$

39. $\cot \beta - \sin \beta = (\cos^2 \beta + \cos \beta - 1)\csc \beta$

40. $(\tan 2\beta - \cot 2\beta)^2 = \tan^2 2\beta(2 - \csc^2 2\beta)^2$

41. $(1 - \sin^2 \alpha)(1 + \tan^2 \alpha) = 1$

42. $(1 - \cos^2 2\alpha)(1 + \cot^2 2\alpha) = 1$

43. $\dfrac{\cos^2 \gamma + \tan^2 \gamma - 1}{\sin^2 \gamma} = \tan^2 \gamma$

44. $\cot^2 \gamma = \dfrac{\sin^2 \gamma + \cot^2 \gamma - 1}{\cos^2 \gamma}$

45. $(\cos \alpha - \cos \beta)^2 + (\sin \alpha - \sin \beta)^2 = 2 - 2(\cos \alpha \cos \beta + \sin \alpha \sin \beta)$

46. $(\sec \alpha + \sec \beta)^2 - (\tan \alpha - \tan \beta)^2 = 2 + 2(\sec \alpha \sec \beta + \tan \alpha \tan \beta)$

47. $\tan A + \cot B = (\sin A \sin B + \cos A \cos B)\sec A \csc B$

48. $\sec A + \csc B = (\cos A \sin^2 B + \sin B \cos^2 A)\sec^2 A \csc^2 B$

49. $(\sin A \cos A \cos B + \sin B \cos B \cos A)\sec A \sec B = \sin A + \sin B$

50. $(\cos A \cos B \tan A + \sin A \sin B \cot B)\csc A \sec B = 2$

Mind Boggler

51. a. What is an identity?
    b. Are the fundamental identities always true? If not, give examples of angles for which they are not true.
    c. Are the identities in Problems 1, 21, 33, and 37 always true? If not, give angles for which they are not true.

3.3 PROVING IDENTITIES *(CONTINUED)*
(Tricks of the Trade)

There are many "tricks of the trade" that can be used in proving identities. Some of these "tricks of the trade" are developed in the following examples.

EXAMPLE: Prove that $\cos \theta/(1 - \sin \theta) = (1 + \sin \theta)/\cos \theta$.

Solution: Sometimes, when there is a binomial in the numerator or denominator, the identity can be proved by multiplying one side by 1, where 1 is written in the form of the conjugate of the binomial. When changing one side, keep a sharp eye on the other side, since it often gives a clue about what to do. Thus in this example we can multiply the numerator and denominator of the left-hand side by $(1 + \sin \theta)$:

$$\frac{\cos \theta}{1 - \sin \theta} = \frac{\cos \theta}{1 - \sin \theta} \cdot \frac{1 + \sin \theta}{1 + \sin \theta}$$

$$= \frac{\cos \theta(1 + \sin \theta)}{1 - \sin^2 \theta}$$

$$= \frac{\cos \theta(1 + \sin \theta)}{\cos^2 \theta}$$

$$= \frac{1 + \sin \theta}{\cos \theta}$$

We could also have proved this identity by multiplying the numerator and denominator of the right-hand side by $(1 - \sin \theta)$.

EXAMPLE: Prove that $(\sec^2 2\theta - \tan^2 2\theta)/(\tan 2\theta + \sec 2\theta) = \cos 2\theta/(1 + \sin 2\theta)$.

Solution: Sometimes the identity can be proved by factoring.

$$\frac{\sec^2 2\theta - \tan^2 2\theta}{\sec 2\theta + \tan 2\theta} = \frac{(\sec 2\theta + \tan 2\theta)(\sec 2\theta - \tan 2\theta)}{\sec 2\theta + \tan 2\theta}$$

$$= \sec 2\theta - \tan 2\theta$$

$$= \frac{1}{\cos 2\theta} - \frac{\sin 2\theta}{\cos 2\theta}$$

$$= \frac{1 - \sin 2\theta}{\cos 2\theta}$$

$$= \frac{1 - \sin 2\theta}{\cos 2\theta} \cdot \frac{1 + \sin 2\theta}{1 + \sin 2\theta}$$

$$= \frac{1 - \sin^2 2\theta}{\cos 2\theta(1 + \sin 2\theta)}$$

$$= \frac{\cos^2 2\theta}{\cos 2\theta(1 + \sin 2\theta)}$$

$$= \frac{\cos 2\theta}{1 + \sin 2\theta}$$

EXAMPLE: Prove that $(-2 \sin \theta \cos \theta)/(1 - \sin \theta - \cos \theta) = 1 + \sin \theta + \cos \theta$.

Solution: Sometimes, when there is a fraction on one side, the identity can be proved by multiplying the other side by 1 written so that the desired denominator is obtained. Thus for this example

$$1 + \sin \theta + \cos \theta = (1 + \sin \theta + \cos \theta) \cdot \frac{1 - \sin \theta - \cos \theta}{1 - \sin \theta - \cos \theta}$$

$$= \frac{(1 + \sin \theta + \cos \theta)(1 - \sin \theta - \cos \theta)}{1 - \sin \theta - \cos \theta}$$

$$= \frac{1 - \sin \theta - \cos \theta + \sin \theta - \sin^2 \theta - \sin \theta \cos \theta + \cos \theta - \cos \theta \sin \theta - \cos^2 \theta}{1 - \sin \theta - \cos \theta}$$

$$= \frac{1 - (\sin^2 \theta + \cos^2 \theta) - 2 \sin \theta \cos \theta}{1 - \sin \theta - \cos \theta}$$

$$= \frac{-2 \sin \theta \cos \theta}{1 - \sin \theta - \cos \theta}.$$

In summary, there is no one best way to proceed in proving identities. However, the following hints should help.

Tricks of the Trade

1. If one side contains one function only, write all the trigonometric functions on the other side in terms of that function.
2. If the denominator of a fraction consists of only one function, break up the fraction.
3. Simplify by combining fractions.
4. Factoring is sometimes helpful.
5. Change all trigonometric functions to sines and cosines and simplify.
6. Multiply by the conjugate of either the numerator or the denominator.
7. Avoid the introduction of radicals.

PROBLEM SET 3.3

A Problems

Prove the following identities.

1. $\sin \theta = \sin^3 \theta + \cos^2 \theta \sin \theta$

2. $\sec \theta = \sec \theta \sin^2 \theta + \cos \theta$

3. $\tan \theta = \cot \theta \tan^2 \theta$

4. $\dfrac{\sin \theta \cos \theta + \sin^2 \theta}{\sin \theta} = \cos \theta + \sin \theta$

5. $\tan^2 \theta - \sin^2 \theta = \tan^2 \theta \sin^2 \theta$

6. $\cot^2 \theta \cos^2 \theta = \cot^2 \theta - \cos^2 \theta$

7. $\tan A + \cot A = \sec A \csc A$

8. $\cot A = \csc A \sec A - \tan A$

9. $\sin x + \cos x = \dfrac{\sec x + \csc x}{\csc x \sec x}$

10. $\dfrac{\cos \gamma + \tan \gamma \sin \gamma}{\sec \gamma} = 1$

11. $\dfrac{1 - \sec^2 t}{\sec^2 t} = -\sin^2 t$

12. $\dfrac{1 + \cot^2 t}{\cot^2 t} = \sec^2 t$

13. $(\sec \theta - \cos \theta)^2 = \tan^2 \theta - \sin^2 \theta$

14. $\dfrac{\sin \theta}{\csc \theta} + \dfrac{\cos \theta}{\sec \theta} = 1$

15. $1 - \sin 2\theta = \dfrac{1 - \sin^2 2\theta}{1 + \sin 2\theta}$

16. $\dfrac{1 - \tan^2 3\theta}{1 - \tan 3\theta} = 1 + \tan 3\theta$

17. $\sin \lambda = \dfrac{\sin^2 \lambda + \sin \lambda \cos \lambda + \sin \lambda}{\sin \lambda + \cos \lambda + 1}$

18. $\dfrac{1 + \cot 2\lambda \sec 2\lambda}{\tan 2\lambda + \sec 2\lambda} = \cot 2\lambda$

19. $\sin 2\alpha \cos 2\alpha (\tan 2\alpha + \cot 2\alpha) = 1$

20. $(\sin \beta - \cos \beta)^2 + (\sin \beta + \cos \beta)^2 = 2$

21. $\csc 3\beta - \cos 3\beta \cot 3\beta = \sin 3\beta$

22. $\dfrac{1 + \cot^2 A}{1 + \tan^2 A} = \cot^2 A$

23. $\dfrac{\sin^2 B - \cos^2 B}{\sin B + \cos B} = \sin B - \cos B$

24. $\dfrac{\tan^2 \gamma - \cot^2 \gamma}{\tan \gamma + \cot \gamma} = \tan \gamma - \cot \gamma$

25. $\tan^2 2\gamma + \sin^2 2\gamma + \cos^2 2\gamma = \sec^2 2\gamma$

26. $\cot^2 C + \cos^2 C + \sin^2 C = \csc^2 C$

27. $\dfrac{\tan \theta + \cot \theta}{\sec \theta \csc \theta} = 1$

28. $\dfrac{\tan \theta - \cot \theta}{\sec \theta \csc \theta} = \sin^2 \theta - \cos^2 \theta$

29. $\dfrac{1}{\sin \theta + \cos \theta} + \dfrac{1}{\sin \theta - \cos \theta} = \dfrac{\sin \theta}{\sin^2 \theta - \frac{1}{2}}$

30. $\dfrac{1}{\sec \theta + \tan \theta} + \dfrac{1}{\sec \theta - \tan \theta} = 2 \sec \theta$

B Problems

31. $\dfrac{1 + \tan C}{1 - \tan C} = \dfrac{\sec^2 C \cos C + 2 \sin C}{2 \cos C - \sec C}$

32. $(\cot x + \csc x)^2 = \dfrac{\sec x + 1}{\sec x - 1}$

33. $\dfrac{\sin^3 x - \cos^3 x}{\sin x - \cos x} = 1 + \sin x \cos x$

34. $\dfrac{\tan^3 t - \cot^3 t}{\tan t - \cot t} = \sec^2 t + \cot^2 t$

35. $\sqrt{(3 \cos \theta - 4 \sin \theta)^2 + (3 \sin \theta + 4 \cos \theta)^2} = 5$

36. $\dfrac{1 - \cos \theta}{1 + \cos \theta} = \left(\dfrac{1 - \cos \theta}{\sin \theta} \right)^2$

37. $\dfrac{(\sec^2 \gamma + \tan^2 \gamma)^2}{\sec^4 \gamma - \tan^4 \gamma} = 1 + 2 \tan^2 \gamma$

38. $\dfrac{(\cos^2 \gamma - \sin^2 \gamma)^2}{\cos^4 \gamma - \sin^4 \gamma} = 2 \cos^2 \gamma - 1$

39. $(\sec 2\theta + \csc 2\theta)^2 = \dfrac{1 + 2 \sin 2\theta \cos 2\theta}{\cos^2 2\theta \sin^2 2\theta}$

40. $\dfrac{1}{\sec \theta + \tan \theta} = \sec \theta - \tan \theta$

41. $\csc \theta + \cot \theta = \dfrac{1}{\csc \theta - \cot \theta}$

42. $\sec^2 \lambda - \csc^2 \lambda = (2 \sin^2 \lambda - 1)(\sec^2 \lambda + \csc^2 \lambda)$

43. $2 \csc A = 2 \csc A - \cot A \cos A + \cos^2 A \csc A$

44. $\sec^2 2\lambda + \csc^2 2\lambda = \csc^2 2\lambda \sec^2 2\lambda$

45. $\dfrac{\tan \theta}{\cot \theta} - \dfrac{\cot \theta}{\tan \theta} = \sec^2 \theta - \csc^2 \theta$

46. $\dfrac{\cos^4 \theta - \sin^4 \theta}{(\cos^2 \theta - \sin^2 \theta)^2} = \dfrac{\cos \theta}{\cos \theta + \sin \theta} + \dfrac{\sin \theta}{\cos \theta - \sin \theta}$

47. $\dfrac{1 + \tan^3 \theta}{1 + \tan \theta} = \sec^2 \theta - \tan \theta$

48. $\dfrac{1 - \sec^3 \theta}{1 - \sec \theta} = \tan^2 \theta + \sec \theta + 2$

49. $\dfrac{\cos^2 \theta - \cos \theta \csc \theta}{\cos^2 \theta \csc \theta - \cos \theta \csc^2 \theta} = \sin \theta$

50. $\dfrac{\tan^2 \theta - 2 \tan \theta}{2 \tan \theta - 4} = \dfrac{1}{2} \tan \theta$

51. $\dfrac{\cos \theta + \cos^2 \theta}{\cos \theta + 1} = \dfrac{\cos \theta \sin \theta + \cos^2 \theta}{\sin \theta + \cos \theta}$

52. $\dfrac{\csc \theta + 1}{\cot^2 \theta + \csc \theta + 1} = \dfrac{\sin^2 \theta + \sin \theta \cos \theta}{\sin \theta + \cos \theta}$

53. $\sin \theta + \cos \theta + 1 = \dfrac{2 \sin \theta \cos \theta}{\sin \theta + \cos \theta - 1}$

54. $\dfrac{2 \tan^2 \theta + 2 \tan \theta \sec \theta}{\tan \theta + \sec \theta - 1} = \tan \theta + \sec \theta + 1$

55. $\dfrac{\csc \theta + 1}{\csc \theta - 1} - \dfrac{\sec \theta - \tan \theta}{\sec \theta + \tan \theta} = 4 \tan \theta \sec \theta$

56. $\dfrac{\cos\theta + \sin\theta}{\cos\theta - \sin\theta} + \dfrac{\cot\theta - 1}{\cot\theta + 1} = \dfrac{-2}{\sin^2\theta - \cos^2\theta}$

57. $\dfrac{\cos\theta + 1}{\cos\theta - 1} + \dfrac{1 - \sec\theta}{1 + \sec\theta} = -2\cot^2\theta - 2\csc^2\theta$

Mind Boggler

58. Prove that $\sin\theta/(1 - \cos\theta) + \cos\theta/(1 - \sin\theta) = (\sin\theta + \cos\theta + 1)(\sec\theta\,\csc\theta)$.

3.4 ADDITION LAWS *(It All Adds Up)*

When we are proving identities, it is sometimes necessary to simplify the functional value of the sum or difference of two angles. That is, if α and β represent any two angles, we know that in general

$$\cos(\alpha - \beta) \neq \cos\alpha - \cos\beta.$$

In fact, we will now show that

$$\cos(\alpha - \beta) = \cos\alpha\cos\beta + \sin\alpha\sin\beta.$$

This result not only will simplify the cosine of the difference of two angles but also will provide the cornerstone upon which we'll build a great many additional identities.

We begin by finding the length of any chord (in a unit circle) whose corresponding arc is subtended by the central angle θ, where θ is in standard position. Let A be the point $(1, 0)$ and P the point on the intersection of the terminal side of angle θ and the unit circle. This means that the coordinates of P are $(\cos\theta, \sin\theta)$. We now find the length of the chord AP (see Figure 3.1) by using the distance formula:

$$\begin{aligned}
AP &= \sqrt{(1 - \cos\theta)^2 + (0 - \sin\theta)^2} \\
&= \sqrt{1 - 2\cos\theta + \cos^2\theta + \sin^2\theta} \\
&= \sqrt{1 - 2\cos\theta + 1} \\
&= \sqrt{2 - 2\cos\theta}.
\end{aligned}$$

We can now apply this result to a chord determined by any two angles α and β, as shown in Figure 3.2. Let P_α and P_β be the points on the unit circle deter-

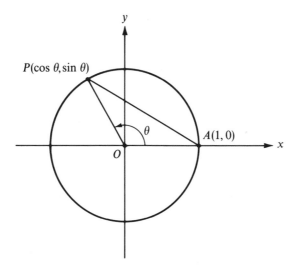

FIGURE 3.1   Length of a chord subtended by an angle θ.

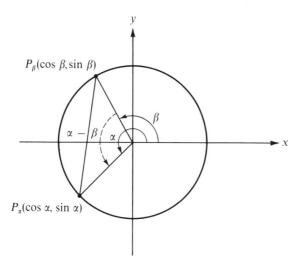

FIGURE 3.2   Distance between P_α and P_β.

mined by the angles α and β, respectively. By the previous result, we see that

$$P_\alpha P_\beta = \sqrt{2 - 2\cos(\alpha - \beta)}.$$

But we could also have found this distance directly via the distance formula.

$$P_\alpha P_\beta = \sqrt{(\cos\beta - \cos\alpha)^2 + (\sin\beta - \sin\alpha)^2}$$

$$= \sqrt{\cos^2\beta - 2\cos\alpha\cos\beta + \cos^2\alpha + \sin^2\beta - 2\sin\alpha\sin\beta + \sin^2\alpha}$$

$$= \sqrt{(\cos^2\beta + \sin^2\beta) + (\cos^2\alpha + \sin^2\alpha) - 2(\cos\alpha\cos\beta + \sin\alpha\sin\beta)}$$

$$= \sqrt{2 - 2(\cos\alpha\cos\beta + \sin\alpha\sin\beta)}$$

Finally, we equate these quantities, since they both represent the distance between P_α and P_β.

$$\sqrt{2 - 2\cos(\alpha - \beta)} = \sqrt{2 - 2(\cos\alpha\cos\beta + \sin\alpha\sin\beta)}$$

$$2 - 2\cos(\alpha - \beta) = 2 - 2(\cos\alpha\cos\beta + \sin\alpha\sin\beta)$$

$$-2\cos(\alpha - \beta) = -2(\cos\alpha\cos\beta + \sin\alpha\sin\beta)$$

$$\cos(\alpha - \beta) = \cos\alpha\cos\beta + \sin\alpha\sin\beta$$

We can use this identity as follows to find the exact value of certain angles.

EXAMPLE: $\cos 345° = \cos 15°$ Reduction principle

$\qquad = \cos(45° - 30°)$ Since $45° - 30° = 15°$

$\qquad = \cos 45° \cos 30° + \sin 45° \sin 30°$ Applying our new identity

$\qquad = \frac{1}{2}\sqrt{2} \cdot \frac{1}{2}\sqrt{3} + \frac{1}{2}\sqrt{2} \cdot \frac{1}{2}$ Using exact values

$\qquad = \dfrac{\sqrt{6}}{4} + \dfrac{\sqrt{2}}{4}$

$\qquad = \dfrac{\sqrt{6} + \sqrt{2}}{4}$

Even though this identity is helpful for making evaluations (as in the preceding example), its real value lies in the fact that it is true for *any* choice of α and β. By making some particular choices for α and β, we find several useful special cases of this identity.

GIVEN: $\cos(\alpha - \beta) = \cos\alpha\cos\beta + \sin\alpha\sin\beta$

1. Let $\alpha = \pi/2$ and $\beta = 0$.

$$\cos(\pi/2 - \theta) = \cos\pi/2\cos\theta + \sin\pi/2\sin\theta$$

$$= 0 \cdot \cos\theta + 1 \cdot \sin\theta$$

$$\therefore \cos(\pi/2 - \theta) = \sin\theta$$

2. Let $\alpha = \pi/2$ and $\beta = \pi/2 - \theta$.

$$\cos[\pi/2 - (\pi/2 - \theta)] = \cos\theta$$

Also, using the identity from part (1),

$$\cos[\pi/2 - (\pi/2 - \theta)] = \sin(\pi/2 - \theta);$$
$$\therefore \sin(\pi/2 - \theta) = \cos\theta.$$

3. $\tan(\pi/2 - \theta) = \dfrac{\sin(\pi/2 - \theta)}{\cos(\pi/2 - \theta)}$

$$= \dfrac{\cos\theta}{\sin\theta}$$

$$= \cot\theta$$

The first three identities tell us that the cofunctions of complementary angles are equal. That is, for any angle θ:

$$\cos(\pi/2 - \theta) = \sin\theta;$$
$$\sin(\pi/2 - \theta) = \cos\theta;$$
$$\tan(\pi/2 - \theta) = \cot\theta.$$

---

4. Let $\alpha = 0$ and $\beta = \theta$.

$$\cos(0 - \theta) = \cos(-\theta)$$

Also,

$$\cos(0 - \theta) = \cos 0 \cos\theta + \sin 0 \sin\theta$$
$$= 1 \cdot \cos\theta + 0 \cdot \sin\theta$$
$$= \cos\theta;$$
$$\therefore \cos(-\theta) = \cos\theta.$$

5. Let $\alpha = \pi/2$ and $\beta = -\theta$

$$\cos(\pi/2 + \theta) = \cos[\pi/2 - (-\theta)]$$
$$= \cos\pi/2 \cos(-\theta) + \sin\pi/2 \sin(-\theta)$$
$$= 0 \cdot \cos(-\theta) + 1 \cdot \sin(-\theta)$$
$$= \sin(-\theta)$$

Also,

$$\cos(\pi/2 + \theta) = \cos(\theta + \pi/2)$$
$$= \cos[\theta - (-\pi/2)]$$
$$= \cos\theta \cos(-\pi/2) + \sin\theta \sin(-\pi/2)$$
$$= \cos\theta \cdot 0 \qquad + \sin\theta \cdot (-1)$$
$$= -\sin\theta;$$
$$\therefore \sin(-\theta) = -\sin\theta.$$

6. $\tan(-\theta) = \dfrac{\sin(-\theta)}{\cos(-\theta)}$

$ = \dfrac{-\sin\theta}{\cos\theta}$

$ = -\tan\theta$

The second three identities help us deal with a trigonometric function of a negative angle.

For any angle θ:

$$\cos(-\theta) = \cos\theta;$$
$$\sin(-\theta) = -\sin\theta;$$
$$\tan(-\theta) = -\tan\theta.$$

---

7. $\cos(\alpha + \beta) = \cos[\alpha - (-\beta)]$

$ = \cos\alpha\cos(-\beta) + \sin\alpha\sin(-\beta)$

$ = \cos\alpha\cos\beta - \sin\alpha\sin\beta$

8. $\sin(\alpha + \beta) = \cos[\pi/2 - (\alpha + \beta)]$

$ = \cos[(\pi/2 - \alpha) - \beta]$

$ = \cos(\pi/2 - \alpha)\cos\beta + \sin(\pi/2 - \alpha)\sin\beta$

$ = \sin\alpha\cos\beta + \cos\alpha\sin\beta$

9. Replace β by $-\beta$ in identity (8) to obtain

$\sin(\alpha - \beta) = \sin\alpha\cos\beta - \cos\alpha\sin\beta$.

Identities 7–9, together with the original given identity, are called the fundamental *sum and difference identities* and should be memorized.

$$\cos(\alpha + \beta) = \cos\alpha\cos\beta - \sin\alpha\sin\beta$$
$$\cos(\alpha - \beta) = \cos\alpha\cos\beta + \sin\alpha\sin\beta$$
$$\sin(\alpha + \beta) = \sin\alpha\cos\beta + \cos\alpha\sin\beta$$
$$\sin(\alpha - \beta) = \sin\alpha\cos\beta - \cos\alpha\sin\beta$$
$$\tan(\alpha + \beta) = \frac{\tan\alpha + \tan\beta}{1 - \tan\alpha\tan\beta}$$
$$\tan(\alpha - \beta) = \frac{\tan\alpha - \tan\beta}{1 + \tan\alpha\tan\beta}$$

The proof of this last identity, involving tangents, is left as a problem. At the opening of Chapter 2 you read a little about Ptolemy's role in trigonometry. The identity for the sine of the sum and difference of two angles is sometimes called Ptolemy's Theorem.

PROBLEM SET 3.4

A Problems

Using the identities of this section, find the exact values of the sine, cosine, and tangent of the angles given in Problems 1–10.

1. $-15°$
2. $195°$
3. $75°$
4. $165°$
5. $345°$
6. $105°$
7. $225°$
8. $255°$
9. $285°$
10. $-75°$

Change each of the expressions in Problems 11–19 to a function of θ only.

11. $\cos(30° + \theta)$
12. $\sin(\theta - 45°)$
13. $\tan(\pi/4 + \theta)$
14. $\cos(\theta - 45°)$
15. $\sin(120° + \theta)$
16. $\tan(\theta - 225°)$
17. $\cos(\theta + \theta)$
18. $\sin(\theta + \theta)$
19. $\tan(\theta + \theta)$

B Problems

Evaluate the expressions in Problems 20–25. You may use Table III or a calculator.

20. $\sin 158° \cos 92° - \cos 158° \sin 92°$
21. $\cos 114° \cos 85° + \sin 114° \sin 85°$
22. $\cos 30° \cos 48° - \sin 30° \sin 48°$
23. $\sin 18° \cos 23° + \cos 18° \sin 23°$
24. $\dfrac{\tan 32° + \tan 18°}{1 - \tan 32° \tan 18°}$
25. $\dfrac{\tan 59° - \tan 25°}{1 + \tan 59° \tan 25°}$

26. Prove that $\tan(\alpha + \beta) = (\tan \alpha + \tan \beta)/(1 - \tan \alpha \tan \beta)$.
27. Prove that $\tan(\alpha - \beta) = (\tan \alpha - \tan \beta)/(1 + \tan \alpha \tan \beta)$.
28. Derive a formula for $\cot(\alpha + \beta)$ in terms of $\cot \alpha$ and $\cot \beta$.
29. Derive a formula for $\cot(\alpha - \beta)$ in terms of $\cot \alpha$ and $\cot \beta$.

In Problems 30–33 let $\sin \alpha = 3/5$ and $\sin \beta = 5/13$, where α and β are both acute. Find the given value.

EXAMPLE: $\sin(\alpha - \beta)$.

Solution: $\sin(\alpha - \beta) = \sin \alpha \cos \beta - \cos \alpha \sin \beta$

$$= \left(\frac{3}{5}\right) \cos \beta - \cos \alpha \left(\frac{5}{13}\right)$$

since $\sin \alpha$ and $\sin \beta$ are given. We find the other values by using fundamental identities:

$$\cos \beta = \pm\sqrt{1 - \sin^2 \beta} \qquad\qquad \cos \alpha = \pm\sqrt{1 - \sin^2 \alpha}$$

$$= \pm\sqrt{1 - (5/13)^2} \qquad\qquad = \pm\sqrt{1 - (3/5)^2}$$

$$= \pm\sqrt{144/169} \qquad\qquad = \pm\sqrt{16/25}$$

$$= 12/13 \qquad\qquad\qquad = 4/5$$

where the positive values are chosen since it is given that α and β are both acute. We now can finish the problem.

$$\sin(\alpha - \beta) = \left(\frac{3}{5}\right)\left(\frac{12}{13}\right) - \left(\frac{4}{5}\right)\left(\frac{5}{13}\right)$$

$$= \frac{36}{65} - \frac{20}{65}$$

$$= \frac{16}{65}$$

30. $\cos(\alpha + \beta)$ 31. $\cos(\alpha - \beta)$ 32. $\sin(\alpha + \beta)$ 33. $\tan(\alpha + \beta)$

In Problems 34–37 let $\sin \alpha = 8/17$ and $\tan \beta = 9/40$, where α is in quadrant II and β is in quadrant III. Find the given value.

34. $\sin(\alpha + \beta)$ 35. $\sin(\alpha - \beta)$ 36. $\cos(\alpha - \beta)$ 37. $\tan(\alpha - \beta)$

Prove the identities in Problems 38–48.

38. $\dfrac{\cos 5\theta}{\sin \theta} - \dfrac{\sin 5\theta}{\cos \theta} = \dfrac{\cos 6\theta}{\sin \theta \cos \theta}$

39. $\dfrac{\sin 6\theta}{\sin 3\theta} - \dfrac{\cos 6\theta}{\cos 3\theta} = \sec 3\theta$

40. $\sin(\alpha + \beta) \cos \beta - \cos(\alpha + \beta) \sin \beta = \sin \alpha$

41. $\cos(\alpha - \beta) \cos \beta - \sin(\alpha - \beta) \sin \beta = \cos \alpha$

42. $\dfrac{\tan(\alpha + \beta) - \tan \beta}{1 + \tan(\alpha + \beta) \tan \beta} = \tan \alpha$

43. $\dfrac{\sin(\theta + h) - \sin \theta}{h} = \cos \theta\left(\dfrac{\sin h}{h}\right) - \sin \theta\left(\dfrac{1 - \cos h}{h}\right)$

44. $\dfrac{\cos(\theta + h) - \cos \theta}{h} = -\sin \theta\left(\dfrac{\sin h}{h}\right) - \cos \theta\left(\dfrac{1 - \cos h}{h}\right)$

45. $\sin(\alpha + \beta + \gamma) = \sin \alpha \cos \beta \cos \gamma + \cos \alpha \sin \beta \cos \gamma$
$\qquad\qquad + \cos \alpha \cos \beta \sin \gamma - \sin \alpha \sin \beta \sin \gamma$

46. $\cos(\alpha + \beta + \gamma) = \cos \alpha \cos \beta \cos \gamma - \cos \alpha \sin \beta \sin \gamma$
$\qquad\qquad - \sin \alpha \cos \beta \sin \gamma - \sin \alpha \sin \beta \cos \gamma$

47. $\tan(\alpha + \beta + \gamma) = \dfrac{\tan \alpha + \tan \beta + \tan \gamma - \tan \alpha \tan \beta \tan \gamma}{1 - \tan \beta \tan \gamma - \tan \alpha \tan \gamma - \tan \alpha \tan \beta}$

Mind Bogglers

48. $\cot(\alpha + \beta + \gamma) = \dfrac{\cot \alpha \cot \beta \cot \gamma - \cot \alpha - \cot \beta - \cot \gamma}{\cot \beta \cot \gamma + \cot \alpha \cot \gamma + \cot \alpha \cot \beta - 1}$

 (Hint: First do Problem 28.)

49. Use the identities of this section to show the following.

 a. $\cos 2\theta = \cos^2 \theta - \sin^2 \theta$

 b. $\sin \tfrac{1}{2}\theta = \pm \sqrt{\dfrac{1 - \cos \theta}{2}}$

50. Here is an alternate proof for the formula $\cos(\alpha + \beta) = \cos \alpha \cos \beta - \sin \alpha \sin \beta$.
 a. case i: let $\alpha = 0$ and $\beta = 0$. Prove the given identity.
 b. case ii: let $\alpha \ne 0$ and $\beta \ne 0$. Let α be in standard position and β be drawn so that its initial side is along the terminal side of α (see Figure 3.3). Let P_α be an arbitrary point on the terminal side of α, and draw $P_\alpha P_\beta$ perpendicular to OP_α. Draw perpendiculars $P_\alpha S$ and $P_\beta T$ to the x-axis. Draw $P_\alpha Q$ perpendicular to $P_\beta T$. Thus

$$\cos(\alpha + \beta) = \frac{OT}{OP_\beta}.$$

 Now show

$$\cos(\alpha + \beta) = \frac{OS}{OP_\alpha} \cdot \frac{OP_\alpha}{OP_\beta} - \frac{QP_\alpha}{P_\alpha P_\beta} \cdot \frac{P_\alpha P_\beta}{OP_\beta},$$

 and therefore

$$\cos(\alpha + \beta) = \cos \alpha \cos \beta - \sin \alpha \sin \beta.$$

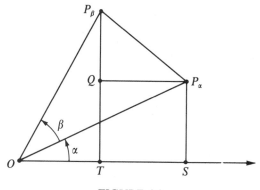

FIGURE 3.3

3.5 DOUBLE-ANGLE AND HALF-ANGLE IDENTITIES (Double the Trouble and Half the Angle)

There are two additional, special cases of the sum and difference identities of the previous section, and they will now be considered. The first is that of the *double-angle identities.* For this case we let $\alpha = \theta$ and $\beta = \theta$.

$$\cos 2\theta = \cos(\theta + \theta) = \cos \theta \cos \theta - \sin \theta \sin \theta$$
$$= \cos^2 \theta - \sin^2 \theta$$
$$\sin 2\theta = \sin(\theta + \theta) = \sin \theta \cos \theta + \cos \theta \sin \theta$$
$$= 2 \sin \theta \cos \theta$$
$$\tan 2\theta = \tan(\theta + \theta) = \frac{\tan \theta + \tan \theta}{1 - \tan \theta \tan \theta}$$
$$= \frac{2 \tan \theta}{1 - \tan^2 \theta}$$

EXAMPLES:

1. $\tan \dfrac{\pi}{8} = \tan\left(2 \cdot \dfrac{\pi}{16}\right) = \dfrac{2 \tan \dfrac{\pi}{16}}{1 - \tan^2 \dfrac{\pi}{16}}$

2. $\cos 100x = \cos(2 \cdot 50x) = \cos^2 50x - \sin^2 50x$

3. $\sin \alpha = 2 \sin \frac{1}{2}\alpha \cos \frac{1}{2}\alpha$

4. $\cos 3\theta = \cos(2\theta + \theta) = \cos 2\theta \cos \theta - \sin 2\theta \sin \theta$
$$= (\cos^2 \theta - \sin^2 \theta)\cos \theta - (2 \sin \theta \cos \theta)\sin \theta$$
$$= \cos^3 \theta - \sin^2 \theta \cos \theta - 2 \sin^2 \theta \cos \theta$$
$$= \cos^3 \theta - (1 - \cos^2 \theta)\cos \theta - 2(1 - \cos^2 \theta)\cos \theta$$
$$= \cos^3 \theta - \cos \theta + \cos^3 \theta - 2 \cos \theta + 2 \cos^3 \theta$$
$$= 4 \cos^3 \theta - 3 \cos \theta$$

Sometimes, as in Example 4, we want to write $\cos 2\theta$ in terms of cosines, and at other times we want to write it in terms of sines. Thus

$$\cos 2\theta = \cos^2 \theta - \sin^2 \theta$$
$$= \cos^2 \theta - (1 - \cos^2 \theta)$$
$$= 2 \cos^2 \theta - 1,$$

and

$$\cos 2\theta = \cos^2 \theta - \sin^2 \theta$$
$$= (1 - \sin^2 \theta) - \sin^2 \theta$$
$$= 1 - 2 \sin^2 \theta.$$

These last two identities lead us to the second important special case of the sum and difference identities, called the *half-angle identities*. We wish to solve

$\cos 2\alpha = 2 \cos^2 \alpha - 1$ for $\cos \alpha$.

$$2 \cos^2 \alpha - 1 = \cos 2\alpha$$

$$2 \cos^2 \alpha = 1 + \cos 2\alpha$$

$$\cos^2 \alpha = \frac{1 + \cos 2\alpha}{2}$$

Now, if we replace $\alpha = \frac{1}{2}\theta$, we see that $2\alpha = \theta$ and

$$\cos^2 \tfrac{1}{2}\theta = \frac{1 + \cos \theta}{2}.$$

If $\frac{1}{2}\theta$ is in quadrant I or IV, then

$$\cos \frac{1}{2}\theta = \sqrt{\frac{1 + \cos \theta}{2}};$$

if $\frac{1}{2}\theta$ is in quadrant II or III, then

$$\cos \frac{1}{2}\theta = -\sqrt{\frac{1 + \cos \theta}{2}}.$$

Sometimes we summarize these results by writing

$$\cos \frac{1}{2}\theta = \pm\sqrt{\frac{1 + \cos \theta}{2}}.$$

However, *we must be careful*. The sign is chosen according to the quadrant of $\frac{1}{2}\theta$. The formula requires either $+$ or $-$, but not both. (This situation is not consistent with the use of $\pm$ from algebra. For example, when using $\pm$ in the quadratic formula, we mean to indicate possibly *two* correct roots. In this trigonometric identity we will obtain *one* correct value depending on $\frac{1}{2}\theta$.)

For the sine we wish to solve $\cos 2\alpha = 1 - 2 \sin^2 \alpha$ for $\sin \alpha$.

$$\cos 2\alpha = 1 - 2 \sin^2 \alpha$$

$$2 \sin^2 \alpha = 1 - \cos 2\alpha$$

$$\sin^2 \alpha = \frac{1 - \cos 2\alpha}{2}$$

We replace $\alpha = \frac{1}{2}\theta$, and

$$\sin^2 \frac{1}{2}\theta = \frac{1 - \cos \theta}{2}$$

or

$$\sin \frac{1}{2}\theta = \pm\sqrt{\frac{1 - \cos \theta}{2}},$$

where the sign depends on the quadrant of $\frac{1}{2}\theta$. If $\frac{1}{2}\theta$ is in quadrant I or II, we use $+$; if it is in quadrant III or IV, we use $-$.

Finally, to find the half-angle identity for the tangent, we write

$$\tan^2 \frac{1}{2}\theta = \frac{\sin^2 \frac{1}{2}\theta}{\cos^2 \frac{1}{2}\theta} = \frac{\dfrac{1 - \cos \theta}{2}}{\dfrac{1 + \cos \theta}{2}} = \frac{1 - \cos \theta}{1 + \cos \theta}.$$

Thus

$$\tan \frac{1}{2}\theta = \pm \sqrt{\frac{1 - \cos \theta}{1 + \cos \theta}}.$$

But this result can be further simplified to

$$\tan \frac{1}{2}\theta = \frac{1 - \cos \theta}{\sin \theta} = \frac{\sin \theta}{1 + \cos \theta}.$$

We now summarize the identities of this section:

Double-Angle Identities	*Half-Angle Identities*
$\cos 2\theta = \cos^2 \theta - \sin^2 \theta$	
$\qquad = 2 \cos^2 \theta - 1$	$\cos \frac{1}{2}\theta = \pm \sqrt{\dfrac{1 + \cos \theta}{2}}$
$\qquad = 1 - 2 \sin^2 \theta$	
$\sin 2\theta = 2 \sin \theta \cos \theta$	$\sin \frac{1}{2}\theta = \pm \sqrt{\dfrac{1 - \cos \theta}{2}}$
$\tan 2\theta = \dfrac{2 \tan \theta}{1 - \tan^2 \theta}$	$\tan \frac{1}{2}\theta = \dfrac{1 - \cos \theta}{\sin \theta}$
	$\qquad = \dfrac{\sin \theta}{1 + \cos \theta}$

EXAMPLES:

1. Find $\cos 9\pi/8$.

    Solution:

    $$\cos 9\pi/8 = \cos \left(\frac{1}{2} \cdot \frac{9\pi}{4} \right) = -\sqrt{\frac{1 + \cos 9\pi/4}{2}}$$

    Choose a negative sign, since $9\pi/8$ is in quadrant III and the cosine is negative in this quadrant.

$$= -\sqrt{\frac{1 + \cos \pi/4}{2}}$$ Note: $\cos 9\pi/4 = \cos \pi/4$
(coterminal angles)

$$= -\sqrt{\frac{1 + \sqrt{2}/2}{2}}$$

$$= -\sqrt{\frac{2 + \sqrt{2}}{4}}$$

$$= -\frac{1}{2}\sqrt{2 + \sqrt{2}}$$

2. If $\cot 2\theta = 3/4$, find $\cos \theta$, $\sin \theta$, and $\tan \theta$ where 2θ is in quadrant I.

Solution: Choose the point $(3, 4)$ on the terminal side of angle 2θ. Then, from the definition of the trigonometric function,

$$\cot 2\theta = \frac{x}{y} \quad \text{and} \quad \cos 2\theta = \frac{x}{\sqrt{x^2 + y^2}}.$$

Thus,

$$\cos 2\theta = \frac{3}{\sqrt{9 + 16}} = \frac{3}{5}.$$

$$\cos \theta = +\sqrt{\frac{1 + 3/5}{2}} \qquad \sin \theta = +\sqrt{\frac{1 - 3/5}{2}}$$

positive value chosen because θ is in quadrant I

$$= \frac{2}{\sqrt{5}} \quad \text{or} \quad \frac{2}{5}\sqrt{5} \qquad = \frac{1}{\sqrt{5}} \quad \text{or} \quad \frac{1}{5}\sqrt{5}$$

$$\tan \theta = \frac{\sin \theta}{\cos \theta} = \frac{1/\sqrt{5}}{2/\sqrt{5}} = \frac{1}{2}$$

3. Prove that $\sin \theta = (2 \tan \tfrac{1}{2}\theta)/(1 + \tan^2 \tfrac{1}{2}\theta)$.

Solution: When proving identities involving functions of different angles, we should write all the trigonometric functions in the problems as functions of a single angle.

As before, we begin with the more complicated side and change to sines and cosines.

$$\frac{2 \tan \tfrac{1}{2}\theta}{1 + \tan^2 \tfrac{1}{2}\theta} = \frac{2\dfrac{\sin \tfrac{1}{2}\theta}{\cos \tfrac{1}{2}\theta}}{1 + \dfrac{\sin^2 \tfrac{1}{2}\theta}{\cos^2 \tfrac{1}{2}\theta}} = \frac{\dfrac{2 \sin \tfrac{1}{2}\theta}{\cos \tfrac{1}{2}\theta}}{\dfrac{\cos^2 \tfrac{1}{2}\theta + \sin^2 \tfrac{1}{2}\theta}{\cos^2 \tfrac{1}{2}\theta}} = \frac{2 \sin \tfrac{1}{2}\theta \cos \tfrac{1}{2}\theta}{\cos^2 \tfrac{1}{2}\theta + \sin^2 \tfrac{1}{2}\theta}$$

$$= 2 \sin \tfrac{1}{2}\theta \cos \tfrac{1}{2}\theta$$

$$= \sin \theta$$

It is sometimes convenient, or even necessary, to write a trigonometric sum as a product or a product as a sum. To do so, we again turn to the identities for the sum and difference of two angles. We add and subtract the following pair of identities:

$$\cos \alpha \cos \beta + \sin \alpha \sin \beta = \cos(\alpha - \beta)$$
$$\cos \alpha \cos \beta - \sin \alpha \sin \beta = \cos(\alpha + \beta)$$

Adding: $2 \cos \alpha \cos \beta = \cos(\alpha - \beta) + \cos(\alpha + \beta)$
Subtracting: $2 \sin \alpha \sin \beta = \cos(\alpha - \beta) - \cos(\alpha + \beta)$

Also adding:

$$\sin \alpha \cos \beta + \cos \alpha \sin \beta = \sin(\alpha + \beta)$$
$$\sin \alpha \cos \beta - \cos \alpha \sin \beta = \sin(\alpha - \beta)$$

$$2 \sin \alpha \cos \beta = \sin(\alpha + \beta) + \sin(\alpha - \beta)$$

These identities are called the *product formulas.*

$$2 \cos \alpha \cos \beta = \cos(\alpha - \beta) + \cos(\alpha + \beta)$$
$$2 \sin \alpha \sin \beta = \cos(\alpha - \beta) - \cos(\alpha + \beta)$$
$$2 \sin \alpha \cos \beta = \sin(\alpha + \beta) + \sin(\alpha - \beta)$$

To return to the original formulas for a sum, we let $x = \alpha + \beta$ and $y = \alpha - \beta$. Solving for α and β, $\alpha = \frac{1}{2}(x + y)$ and $\beta = \frac{1}{2}(x - y)$. Then we substitute into the product formulas to obtain the *sum formulas.*

$$\cos x + \cos y = 2 \cos\left(\frac{x + y}{2}\right) \cos\left(\frac{x - y}{2}\right)$$

$$\cos x - \cos y = -2 \sin\left(\frac{x + y}{2}\right) \sin\left(\frac{x - y}{2}\right)$$

$$\sin x + \sin y = 2 \sin\left(\frac{x + y}{2}\right) \cos\left(\frac{x - y}{2}\right)$$

$$\sin x - \sin y = 2 \sin\left(\frac{x - y}{2}\right) \cos\left(\frac{x + y}{2}\right)$$

(The derivation of the last identity is left as a problem.)

EXAMPLES:

1. Write $\sin 40° \cos 12°$ as the sum of two functions.

   Solution: $2 \sin 40° \cos 12° = \sin(40° + 12°) + \sin(40° - 12°)$
   Therefore, $\sin 40° \cos 12° = \frac{1}{2}(\sin 52° + \sin 28°)$.

2. Write $\sin 35° + \sin 27°$ as a product.

   Solution: $x = 35°$, $y = 27°$, and $(x + y)/2 = (35 + 27)/2 = 31°$; $(x - y)/2 = 4°$.
   Therefore, $\sin 35° + \sin 27° = 2 \sin 31° \cos 4°$.

PROBLEM SET 3.5

A Problems

Use the double-angle or half-angle identities to evaluate Problems 1–10 using exact values.

1. $2 \cos^2 22.5° - 1$

2. $\dfrac{2 \tan \pi/8}{1 - \tan^2 \pi/8}$

3. $\sqrt{\dfrac{1 - \cos 60°}{2}}$

4. $\cos^2 15° - \sin^2 15°$

5. $1 - 2 \sin^2 90°$

6. $-\sqrt{\dfrac{1 - \cos 420°}{2}}$

7. $\sin 22.5°$

8. $\cos \pi/8$

9. $\tan 22.5°$

10. $\sin 105°$

In Problems 11–16 find the exact values of the sine, cosine, and tangent of $\frac{1}{2}\theta$ and 2θ.

11. $\sin \theta = 3/5$; θ in quad I

12. $\sin \theta = 5/13$; θ in quad II

13. $\tan \theta = -5/12$; θ in quad IV

14. $\tan \theta = -3/4$; θ in quad II

15. $\cos \theta = 5/9$; θ in quad I

16. $\cos \theta = -5/13$; θ in quad III

Write each of the expressions in Problems 17–25 as the sum of two functions.

17. $2 \cos 75° \cos 35°$

18. $2 \cos 46° \cos 18°$

19. $2 \sin 35° \sin 24°$

20. $2 \sin 53° \cos 24°$

21. $\sin 70° \sin 88°$

22. $\cos 53° \cos 70°$

23. $\sin 41° \cos 19°$

24. $\sin 2\theta \sin 5\theta$

25. $\cos \theta \cos 3\theta$

Write each of the expressions in Problems 26–33 as a product.

26. $\sin 43° + \sin 63°$

27. $\sin 22° - \sin 6°$

28. $\cos 81° - \cos 79°$

29. $\cos 78° + \cos 25°$

30. $\cos 5x - \cos 3x$

31. $\sin x + \sin 2x$

32. $\sin x - \sin 2x$

33. $\cos 3\theta + \cos 2\theta$

B Problems

In Problems 34–39 find cos θ, sin θ and tan θ when θ is in quadrant I and cot 2θ is given.

34. $\cot 2\theta = -3/4$ 35. $\cot 2\theta = 0$ 36. $\cot 2\theta = 1/\sqrt{3}$

37. $\cot 2\theta = -1/\sqrt{3}$ 38. $\cot 2\theta = -4/3$ 39. $\cot 2\theta = 4/3$

Prove the identities in Problems 40–56.

40. $\tan \dfrac{1}{2}\theta = \dfrac{1 - \cos \theta}{\sin \theta}$ 41. $\tan \dfrac{1}{2}\theta = \dfrac{\sin \theta}{1 + \cos \theta}$

42. $\sin x - \sin y = 2 \sin\left(\dfrac{x - y}{2}\right) \cos\left(\dfrac{x + y}{2}\right)$

43. $\sin \alpha = 2 \sin \alpha/2 \cos \alpha/2$ 44. $\cos 4\theta = \cos^2 2\theta - \sin^2 2\theta$

45. $\sin 2\theta = \dfrac{2 \tan \theta}{1 + \tan^2 \theta}$ 46. $\tan \dfrac{3}{2}\beta = \dfrac{2 \tan(3\beta/4)}{1 - \tan^2(3\beta/4)}$

47. $\dfrac{\sin 5\theta + \sin 3\theta}{\cos 5\theta + \cos 3\theta} = \tan 4\theta$ 48. $\dfrac{\cos 3\theta - \cos \theta}{\sin \theta - \sin 3\theta} = \tan 2\theta$

49. $\csc 2x = \frac{1}{2}\cot x + \frac{1}{2}\tan x$ 50. $\cos 2y = 2 \cos^2 y - 1$

51. $\sin 3\theta = 3 \sin \theta - 4 \sin^3 \theta$ 52. $\tan B/2 = \csc B - \cot B$

53. $\cot 2x = \dfrac{\cot^2 x - 1}{2 \cot x}$ 54. $\sin 4\theta = 4 \sin \theta \cos \theta - 8 \sin^3 \theta \cos \theta$

55. $\tan 3w = \dfrac{3 \tan w - \tan^3 w}{1 - 3 \tan^2 w}$ 56. $\frac{1}{2}\cot x - \frac{1}{2}\tan x = \cot 2x$

Mind Bogglers

57. Prove that $\cos^4 \theta = \frac{1}{8}(3 + 4 \cos 2\theta + \cos 4\theta)$.

58. Prove that $\left(\sqrt{6} + \sqrt{2}\right)/2 = \sqrt{2 + \sqrt{3}}$.

3.6 USING IDENTITIES TO SOLVE EQUATIONS
(Sec and Ye Shall Find)

Up to this point in the chapter we've been working with identities. Sometimes, though, we need to solve trigonometric equations. To do so, we use the techniques of algebra coupled with the use of trigonometric identities. Our goal when solving a trigonometric equation is to first solve for $F(x)$, where F is one of the six trigonometric functions. Next we solve for x by using the inverse relations presented in Section 2.5 or by using the table of exact values. However, for convenience we'll seek those solutions in the interval $0 \le x < 2\pi$. This restriction is somewhat subtle in work with multiple angles, as shown by the following examples.

EXAMPLES:

1. $\sin x = 1/2$. From the table of exact values the reference angle is $\pi/6$. Since the sine is positive in quadrants I and II, $x = \pi/6$ and $5\pi/6$. (If you are using a calculator, find $x = \sin^{-1}(1/2)$. Be sure to give both solutions.)

2. $\sin 2x = 1/2$. Since $0 \leq x < 2\pi$, we solve for $2x$ such that

$$0 \leq 2x < 4\pi.$$

    Thus, $2x = \pi/6, 2x = 5\pi/6, 2x = 13\pi/6$, and $2x = 17\pi/6$; $x = \pi/12, 5\pi/12, 13\pi/12$, and $17\pi/12$.

Notice from the last example that $F(nx)$ requires that we find the values $0 \leq nx < 2n\pi$ for a given trigonometric function F. This means that $0 \leq x < 2\pi$.

We may have to solve for $F(x)$ algebraically by factoring or by using the quadratic formula, as shown by the next examples.

EXAMPLES:

1. $2 \cos \theta \sin \theta = \sin \theta$

Solution: $2 \cos \theta \sin \theta - \sin \theta = 0$

$\quad\quad\quad\quad \sin \theta (2 \cos \theta - 1) = 0$

$\quad\quad\quad\quad \sin \theta = 0 \quad\quad 2 \cos \theta - 1 = 0$

$\quad\quad\quad\quad\quad\quad\quad\quad\quad\quad\quad\quad \cos \theta = 1/2$

$\quad\quad\quad\quad \theta = 0, \pi \quad\quad \theta = \pi/3, 5\pi/3$

$\{0, \pi, \pi/3, 5\pi/3\}$

2. $2 \sin^2 \theta = 1 + 2 \sin \theta$

Solution: $2 \sin^2 \theta - 2 \sin \theta - 1 = 0$

Since we can't factor the left side, we use the quadratic formula to solve for $\sin \theta$.

$$\sin \theta = \frac{2 \pm \sqrt{4 - 4(2)(-1)}}{2(2)}$$

$$= \frac{1 \pm \sqrt{3}}{2}$$

$$\approx 1.366, -0.366025$$

$\quad\quad\quad$ └─ Reject, since $0 \le \sin \theta \le 1$.

Solve $\sin \theta = -0.366025$ by using Table IV or a calculator to find a reference angle of 0.3747. Since the sine is negative in quadrants III and IV, we have 3.5163 and 5.9084. In degrees, $\theta = 201.47°, 338.53°$.

Finally, it may be necessary to apply some trigonometric identities before solving. If different functions appear in the equation, we use the fundamental identities to express the equation in terms of a single function—or at least in terms of factors that contain only a single function. If the *angles* are different (as in θ, 2θ, or 3θ), then we use identities to write the equation in terms of a single angle. If it is necessary to multiply both sides by a trigonometric function or to square both sides, we must be sure to check for extraneous roots.

EXAMPLES:

1. $\sin^2 3\theta + \sin 3\theta = \cos^2 3\theta - 1$

Solution: $\sin^2 3\theta - \cos^2 3\theta + \sin 3\theta + 1 = 0$

$\quad\quad\quad\quad \sin^2 3\theta - (1 - \sin^2 3\theta) + \sin 3\theta + 1 = 0$ $\quad\quad$ Changing to a single function

$\quad\quad\quad\quad 2 \sin^2 3\theta + \sin 3\theta = 0$

$$\sin 3\theta(2 \sin 3\theta + 1) = 0 \qquad\qquad \text{Factoring}$$

$$\sin 3\theta = 0, -1/2$$

$$3\theta = 0, \pi, 2\pi, 3\pi, 4\pi, 5\pi \qquad\qquad \text{From the first factor}$$

$$3\theta = 7\pi/6, 11\pi/6, 19\pi/6, 23\pi/6, 31\pi/6, 35\pi/6 \quad \text{From the second factor}$$
(the reference angle is $\pi/6$ in quadrants III and IV)

$$\{0, \pi/3, 2\pi/3, \pi, 4\pi/3, 5\pi/3, 7\pi/18, 11\pi/18, 19\pi/18, 23\pi/18, 31\pi/18, 35\pi/18\}$$

2. $\cos 6\theta + \sin 3\theta + 1 = 0$

Solution: We notice that the angles 6θ and 3θ are different. We could use a half-angle identity to change $\sin 3\theta$ to a function of 6θ, but this process would introduce a radical. So instead we'll use a double-angle identity to change $\cos 6\theta$ to a function of 3θ as follows:

$$\cos^2 3\theta - \sin^2 3\theta + \sin 3\theta + 1 = 0.$$

The problem is now similar to the one solved in the preceding example.

To summarize the procedure of solving a trigonometric equation, we must carry out three steps.

1. Solve for a trigonometric function. (Algebraic solution, although some trigonometric identities may be required.)
2. Solve for the angle. (Trigonometric solution; may use exact values or inverse functions on a calculator.)
3. Solve for the unknown. (Algebraic solution). The unknown may or may not be the same as the angle. For example, in the preceding example the angle is 3θ and the unknown is θ.

PROBLEM SET 3.6

A Problems

Solve the equations in Problems 1–30 for $0 \le x < 360°$.

1. $\cos x = 1/2$
2. $\cos 2x = 1/2$
3. $\sin 2x = \sqrt{2}/2$
4. $\sin 2x = -\sqrt{3}/2$
5. $\tan 3x = 1$
6. $\sec 2x = -2\sqrt{3}/3$

7. $(\sin x)\cos x = 0$

8. $(\sec 2x)\tan x = 0$

9. $(\sec x - 2)(\sin x - 1) = 0$

10. $(\csc x - 2)(2 \cos x - 1) = 0$

B Problems

11. $\tan^2 x = \sqrt{3} \tan x$

12. $\tan^2 x = \tan x$

13. $\sin x(3 \cos x - 1) = 0$

14. $\sin^2 x = 1$

15. $4 \sin^2 x = 1$

16. $2 \cos 2x \sin 2x = \sin 2x$

17. $\sin x = \cos x$

18. $\sin^2 x = 1/2$

19. $\cos^2 x = 1/2$

20. $\sin 2x = \cos 2x$

21. $\sin^2 x - \sin x - 2 = 0$

22. $\sin^2 x + \cos x = 0$

23. $4 \cos^2 x - 8 \cos x + 3 = 0$

24. $\tan^2 x - 3 \tan x + 1 = 0$

25. $\sec^2 x - \sec x - 1 = 0$

26. $2 \sin^2 x - \cos 2x = 0$

27. $\cos 2x = 3 \sin x$

28. $\sin x = \cos 2x$

29. $4 \sin^3 x + \sin 3x - 3 \sin x + \sqrt{3} = 2 \sin 2x$

30. $4 \cos^3 x - 3 \cos x - \cos 3x + 2 \cos x = 1$

Mind Boggler

31. For $0 \le \theta \le \pi/2$, solve

$$\left(\frac{16}{81}\right)^{\sin^2\theta} + \left(\frac{16}{81}\right)^{\cos^2\theta} = \frac{26}{27}.$$

3.7 SUMMARY AND REVIEW

TERMS

Conditional equation (3.1)   Identity (3.1)
Contradiction (3.1)   Variable equation (3.1)
Equation (3.1)

CONCEPTS *(Summary of Identities)*

(3.1) *Eight Fundamental Identities*

  Reciprocal Identities

1. $\sec \theta = \dfrac{1}{\cos \theta}$ 2. $\csc \theta = \dfrac{1}{\sin \theta}$ 3. $\cot \theta = \dfrac{1}{\tan \theta}$

  Ratio Identities

4. $\tan \theta = \dfrac{\sin \theta}{\cos \theta}$ 5. $\cot \theta = \dfrac{\cos \theta}{\sin \theta}$

  Pythagorean Identities

6. $\sin^2 \theta + \cos^2 \theta = 1$ 7. $1 + \tan^2 \theta = \sec^2 \theta$
8. $1 + \cot^2 \theta = \csc^2 \theta$

(3.4) *Sum and Difference Identities*

9. $\cos(\alpha + \beta) = \cos \alpha \cos \beta - \sin \alpha \sin \beta$
   $\cos(\alpha - \beta) = \cos \alpha \cos \beta + \sin \alpha \sin \beta$

10. $\sin(\alpha + \beta) = \sin \alpha \cos \beta + \cos \alpha \sin \beta$
    $\sin(\alpha - \beta) = \sin \alpha \cos \beta - \cos \alpha \sin \beta$

11. $\tan(\alpha + \beta) = \dfrac{\tan \alpha + \tan \beta}{1 - \tan \alpha \tan \beta}$

    $\tan(\alpha - \beta) = \dfrac{\tan \alpha - \tan \beta}{1 + \tan \alpha \tan \beta}$

(3.5) *Double-Angle Identities*

12. $\cos 2\theta = \cos^2 \theta - \sin^2 \theta$
    $\qquad\quad = 2\cos^2 \theta - 1$
    $\qquad\quad = 1 - 2\sin^2 \theta$

13. $\sin 2\theta = 2 \sin \theta \cos \theta$

14. $\tan 2\theta = \dfrac{2 \tan \theta}{1 - \tan^2 \theta}$

(3.5) *Half-Angle Identities*

$$15. \quad \cos \frac{1}{2}\theta = \pm\sqrt{\frac{1 + \cos \theta}{2}}$$

$$16. \quad \sin \frac{1}{2}\theta = \pm\sqrt{\frac{1 - \cos \theta}{2}}$$

$$17. \quad \tan \frac{1}{2}\theta = \frac{1 - \cos \theta}{\sin \theta}$$

$$= \frac{\sin \theta}{1 + \cos \theta}$$

(3.5) *Product Identities*

18. $2 \cos \alpha \cos \beta = \cos(\alpha - \beta) + \cos(\alpha + \beta)$

19. $2 \sin \alpha \sin \beta = \cos(\alpha - \beta) - \cos(\alpha + \beta)$

20. $2 \sin \alpha \cos \beta = \sin(\alpha + \beta) + \sin(\alpha - \beta)$

(3.5) *Sum Identities*

$$21. \quad \cos \alpha + \cos \beta = 2 \cos\left(\frac{\alpha + \beta}{2}\right) \cos\left(\frac{\alpha - \beta}{2}\right)$$

$$22. \quad \cos \alpha - \cos \beta = -2 \sin\left(\frac{\alpha + \beta}{2}\right) \sin\left(\frac{\alpha - \beta}{2}\right)$$

$$23. \quad \sin \alpha + \sin \beta = 2 \sin\left(\frac{\alpha + \beta}{2}\right) \cos\left(\frac{\alpha - \beta}{2}\right)$$

$$24. \quad \sin \alpha - \sin \beta = 2 \sin\left(\frac{\alpha - \beta}{2}\right) \cos\left(\frac{\alpha + \beta}{2}\right)$$

(3.4) *Miscellaneous Identities*

25. $\cos(\pi/2 - \theta) = \sin \theta$ 26. $\sin(\pi/2 - \theta) = \cos \theta$

27. $\tan(\pi/2 - \theta) = \cot \theta$ 28. $\cos(-\theta) = \cos \theta$

29. $\sin(-\theta) = -\sin \theta$ 30. $\tan(-\theta) = -\tan \theta$

PROBLEM SET 3.7 *(Chapter 3 Test)*

1. (3.1) State from memory and prove the eight fundamental identities.

2. (3.4) Derive an expression for $\tan(\alpha + \beta + \gamma)$.

3. (3.1) Find the other trig functions so that $\sin \theta = 3/5$ when $\tan \theta < 0$.

4. (3.5) Write $\sin(x + h) - \sin x$ as a product.

(3.2 and 3.3) Prove the identities in Problems 5–7.

5. $\dfrac{1 + \tan^2 \theta}{\csc \theta} = \sec \theta \tan \theta$

6. $\dfrac{\cos \theta}{\sec \theta} - \dfrac{\sin \theta}{\cot \theta} = \dfrac{\cos \theta \cot \theta - \tan \theta}{\csc \theta}$

7. $\dfrac{\sin 5\theta + \sin 3\theta}{\cos 5\theta - \cos 3\theta} = -\cot \theta$

(3.6) Solve the equations in Problems 8–10 for $0 \le \theta < 2\pi$.

8. $2 \cos^2 \theta - \sin^2 \theta = 2$

9. $\sin \theta \cos 2\theta - \cos \theta \sin 2\theta = 1$

10. $4 \cos^2 2\theta + 4 \sin 2\theta = 0$

In the previous two chapters we were concerned with the first main use of trigonometry—namely, using the identities to treat the trigonometric functions as important mathematical relationships. In this and the next chapter we are concerned with the second main use of trigonometry—solving triangles to find unknown parts. In this chapter we'll look at right triangles, and in Chapter 5 we'll generalize to include all triangles.

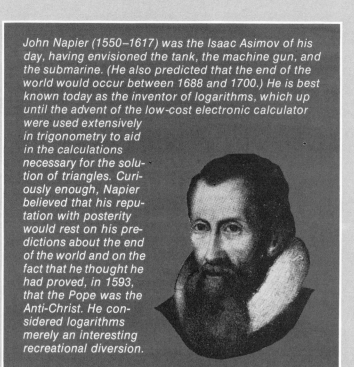

John Napier (1550–1617) was the Isaac Asimov of his day, having envisioned the tank, the machine gun, and the submarine. (He also predicted that the end of the world would occur between 1688 and 1700.) He is best known today as the inventor of logarithms, which up until the advent of the low-cost electronic calculator were used extensively in trigonometry to aid in the calculations necessary for the solution of triangles. Curiously enough, Napier believed that his reputation with posterity would rest on his predictions about the end of the world and on the fact that he thought he had proved, in 1593, that the Pope was the Anti-Christ. He considered logarithms merely an interesting recreational diversion.

4

APPLICATIONS OF TRIGONOMETRY

Right Triangles and Vectors

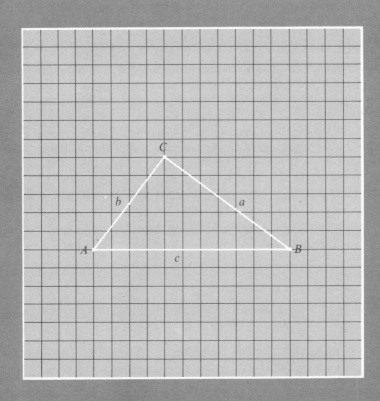

4.1 RIGHT TRIANGLES *(Right On!)*

One of the most important uses of trigonometry is for the solution of triangles. Recall from geometry that every triangle has three sides and three angles, which are called the six *parts* of the triangle. We say that a *triangle is solved* if all six parts are known. Typically, three parts will be given, or known, and it will be our task to find the other three parts. We will generally label a triangle as shown in Figure 4.1.

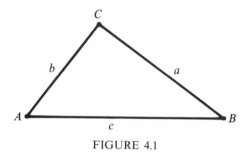

FIGURE 4.1

The vertices will be labeled *A*, *B*, and *C*, with the sides opposite those vertices *a*, *b*, and *c*, respectively. In this chapter we'll limit our examples to right triangles, and we'll further agree that *C* will denote the right angle and *c* the hypotenuse.

According to the definition of the trigonometric functions, the angle under consideration must be in standard position. This requirement is sometimes inconvenient, so we use that definition to give us the following special case, which applies to any acute angle θ of a right triangle. Notice that in Figure 4.1 θ might be angle *A* or angle *B* but would not be angle *C*, since angle *C* is not an acute angle. Also notice that the hypotenuse is one of the sides of both acute angles. The other side making up the angle is called the *adjacent side*. Thus side *a* is adjacent to angle *B*, and side *b* is adjacent to angle *A*. The third side of the triangle (the one not making up the angle) is called the side *opposite* the angle. Thus side *a* is opposite angle *A*, and side *b* is opposite angle *B*.

If θ is an acute angle in a right triangle,

$$\sin \theta = \frac{\text{opposite side}}{\text{hypotenuse}};$$

$$\cos \theta = \frac{\text{adjacent side}}{\text{hypotenuse}};$$

$$\tan \theta = \frac{\text{opposite side}}{\text{adjacent side}}.$$

The other trigonometric functions are the reciprocals of these relationships.

One memory device for remembering these trigonometric functions is

OSCAR HAD A HEAP OF APPLES,

written as follows.

sine: $\dfrac{\text{OSCAR}}{\text{HAD}}$ $\dfrac{\text{O}}{\text{H}}$ $\dfrac{\text{Opp}}{\text{Hyp}}$

cosine: $\dfrac{\text{A}}{\text{HEAP}}$ $\dfrac{\text{A}}{\text{H}}$ $\dfrac{\text{Adj}}{\text{Hyp}}$

tangent: $\dfrac{\text{OF}}{\text{APPLES}}$ $\dfrac{\text{O}}{\text{A}}$ $\dfrac{\text{Opp}}{\text{Adj}}$

We can now use these definitions to solve some given triangles.

EXAMPLES: Indicate the relationships you would use to solve the right triangles ($C = 90°$) with the given information.

1. $a = 50$; $A = 35°$ (Note: when we write $A = 35°$, we mean that the measure of angle A is 35°.)

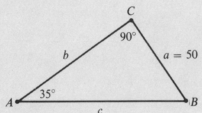

Solution: $A = 35°$ Given

$B = 55°$ Since $A + B = 90°$ for any right triangle with right angle at C.

$C = 90°$ Given

$a = 50$ Given

b: $\tan 35° = \dfrac{50}{b}$

$b = \dfrac{50}{\tan 35°}$

   $= 50 \cot 35°$ (if we wish to avoid division)

c: $\sin 35° = \dfrac{50}{c}$

$c = \dfrac{50}{\sin 35°}$ or $c = 50 \csc 35°$

2. $a = 32; b = 58$

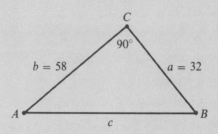

Solution: A: $\tan A = \dfrac{32}{58}$

or $A = \tan^{-1}\left(\dfrac{32}{58}\right)$

B: $\tan B = \dfrac{58}{32}$

or after we've found A we can use $B = 90° - A$

$C = 90°$ Given

$a = 32$ Given

$b = 58$ Given

$c = \sqrt{a^2 + b^2}$ or $\sin A = \dfrac{32}{c}$

$c = \dfrac{32}{\sin A}$ or $32 \csc A$, which would

be used after we've found A.

As you can see, there are many ways of solving a triangle; the method you choose will probably depend on the accuracy you wish and the type of table or calculator you have available.

PROBLEM SET 4.1

A Problems

Indicate the relationships you would use to solve the right triangles ($C = 90°$) with the given information. You do not need to do any arithmetic.

1. $a = 80; B = 60°$

2. $b = 37; A = 69°$

3. $a = 68; b = 83$

4. $a = 29; A = 76°$

5. $b = 13; B = 65°$

6. $a = 69; c = 73$

7. $b = 90; B = 13°$

8. $a = 49; B = 45°$

9. $a = 24; b = 29$ 10. $b = 82; A = 50°$

11. $c = 28.3; A = 69.2°$ 12. $c = 36; A = 6°$

13. $B = 57.4°; a = 70.0$ 14. $A = 56.00°; b = 2350$

15. $B = 23°; a = 9000$ 16. $b = 3100; c = 3500$

17. $B = 16.4°; b = 2580$ 18. $A = 42°; b = 350$

19. $b = 3200; c = 7700$ 20. $b = 4100; c = 4300$

4.2 ACCURACY AND ROUNDING OFF *(Press the Button, Turn the Crank)*

Whenever we work with measurements, the quantities are necessarily ap-proximations. The digits known to be correct in a number obtained by a mea-surement are called *significant digits*. The digits 1, 2, 3, 4, 5, 6, 7, 8, and 9 are always significant, whereas the digit 0 may or may not be significant.

1. Zeros that come between two other digits are significant, as in 203 or 10.04.

2. If the zero's only function is to place the decimal point, it is not significant, as in

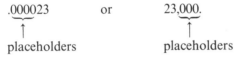

        .000023   or   23,000.

        placeholders   placeholders

If it does more than fix the decimal point, it is significant, as in

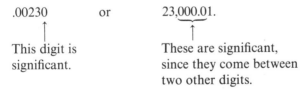

        .00230   or   23,000.01.

        This digit is   These are significant,
        significant.   since they come between
                              two other digits.

This second rule can, of course, result in certain ambiguities, such as in 23,000 (measured to the *exact* unit). To avoid such confusion, we use scientific nota-tion in this case.

        2.3×10^4 has two significant digits.

        2.3000×10^4 has five significant digits.

EXAMPLES:

1. Two significant digits: 46, 0.00083, 4.0×10^1, 0.050

2. Three significant digits: 523, 403, 4.00×10^2, 0.000800

3. Four significant digits: 600.1, 4.000×10^1, 0.0002345

When we are doing calculations with approximate numbers (particularly when using a calculator), it is often necessary to round off results.

To round off numbers:

1. increase the last retained digit by 1 if the residue is larger than 5, or
2. retain the last digit unchanged if the residue is smaller than 5, or
3. retain the last digit unchanged if even, or increase it by 1 if odd, if the residue is exactly 5.

Elaborate rules for computation of approximate data can be developed (when it is necessary for some applications, such as in chemistry), but there are two simple rules that will work satisfactorily for the material of this text.

Addition-Subtraction: Add or subtract in the usual fashion, and then round off the result so that the last digit retained is in the column farthest to the right in which both given numbers have significant digits.

Multiplication-Division: Multiply or divide in the usual fashion, and then round off the result to the smaller number of significant digits found in either of the given numbers.

These agreements are particularly important when we are using a calculator, since the results obtained will look much more accurate on the calculator than they actually are. In Example 1 of Section 4.1 we found

$$b = \frac{50}{\tan 35°}.$$

From Table III, $\tan 35° = .7002$. Thus, $b = 50/.7002 = 71.408169090 \cdots$ (by division), or

$$b = \frac{50}{\tan 35°} = 50 \cot 35°$$

$$= 50(1.428) \qquad \text{(Table III)}$$

$$= 71.400.$$

Or, on a calculator (set to degrees):

Algebraic Format	*Polish Format*

The answer is now displayed:

$$b = \frac{50}{\tan 35°} = 71.40740034.$$

Notice that all the answers above differ. We now use the multiplication-division rule.

50:    This number has one or two significant digits; there is no ambiguity if we write 5×10^1 or 5.0×10^1. *In this book,* if the given data include a number whose degree of accuracy is doubtful, *we will assume the maximum degree of accuracy.* Thus, 50 has two significant digits.

tan 35°:   From Table III we find this number to 4 significant digits; on a calculator you may have 8, 10, or 12 significant digits (depending on the calculator).

The result of this division is correct to two significant digits—namely,

$$b = \frac{50}{\tan 35°} = 71,$$

which agrees with all the above methods of solution.

In solving triangles in this text, we will assume a certain relationship in the accuracy of the measurement between the sides and the angles.

Accuracy in Sides	*Accuracy in Angles*
two significant figures	nearest degree
three significant figures	nearest tenth of a degree
four significant figures	nearest hundredth of a degree

This chart means that, if the data include one side given with two significant digits and another with three significant digits, the angle would be computed to the nearest degree. If one side is given to four significant digits and an angle to the nearest tenth of a degree, then the other sides would be given to three significant digits and the angles computed to the nearest tenth of a degree. In general, results computed from the above table should not be more accurate than the least accurate item of the given data.

If you have access only to a four-function calculator, you can use Table III in conjunction with your calculator. For example, to find b, you first find tan 35° = .7002 and then calculate

$$b = \frac{50}{.7002}$$ $\boxed{5}\,\boxed{0}\,\boxed{\div}\,\boxed{.}\,\boxed{7}\,\boxed{0}\,\boxed{0}\,\boxed{2}\,\boxed{=}$

or

$= 71.41,$ $\boxed{5}\,\boxed{0}\,\boxed{\text{ENTER}}\,\boxed{.}\,\boxed{7}\,\boxed{0}\,\boxed{0}\,\boxed{2}\,\boxed{\div}$

or, to two significant digits, $b = 71$. Of course, one of the great advantages of calculators is that they enable us to work with much greater accuracy without having to use interpolation or tables.

PROBLEM SET 4.2

A Problems

1–20. Solve the problems of Problem Set 4.1.

Problems 21–39 are calculator problems. If you don't have a calculator, you may skip these problems or use logarithms and Table I. Be sure to round off your answer to the appropriate number of significant digits.

21. $(5.813)(6.218)$

22. $4.346 + 25.38 + 4.01 + 2.8736$

23. $42.8 + 36.49 + 0.00123$

24. $(4.83)(9.872) + 6.481$

25. $\dfrac{6.38 \times 4.92}{5.98}$

26. $\dfrac{(12.40)(6.831)(24.86)}{(7.83)(4.86)}$

27. $(4.28)^2(3.24)$

28. $\sqrt{48.63}$

29. $(6.28)^{1/2}(4.85)$

30. $\dfrac{1}{\sqrt{4.83} + \sqrt{2.51}}$

31. $[(4.083)^{0.681}(4.283)^{2.15}]^{-2/3}$

32. $\dfrac{(2.51)^2 + (6.48)^2 - (2.51)(6.48)(0.3462)}{(2.51)(6.48)}$

33. $\dfrac{241^2 + 568^2 - (241)(568)(0.5213)}{(241)(568)}$

34. $\dfrac{23.7 \sin 36.2°}{45.1}$

35. $\dfrac{461 \sin 43.8°}{1215}$

36. $\dfrac{62.8 \sin 81.5°}{\sin 42.3°}$

37. $\dfrac{4.381 \sin 49.86°}{\sin 71.32°}$

38. $\sin^{-1}\left(\dfrac{16 \sin 22°}{25}\right)$

39. $\sin^{-1}\left(\dfrac{42 \sin 52°}{61}\right)$

B Problems

Solve the right triangles in Problems 40–53. Use a calculator or interpolation if necessary.

40. $c = 75.4; \ A = 62.5°$

41. $c = 28.7; \ A = 67.8°$

42. $c = 343.6; \ B = 32.17°$

43. $c = 418.7; \ B = 61.05°$

44. $b = 0.446; \ A = 81.4°$

45. $b = 0.316; \ A = 48.32°$

46. $b = 0.8024; \ B = 24.16°$

47. $b = 0.3596; \ B = 76.06°$

48. $b = 6474; \ A = 17.15°$

49. $b = 34.53; \ c = 73.33$

50. $b = 333; \ a = 106$

51. $b = 95.2; \ a = 159.57$

52. $a = 19.57; \ c = 60.04$

53. $a = 8.85; \ c = 42.78$

54. What is the angle that the diagonal of a 3.0 m by 5.0 m rectangle makes with each of its sides?

55. What is the angle that the diagonal of a level 45 m × 28 m rectangular field makes with each of its sides?

Mind Bogglers

56.   A given triangle *ABC* is *not* a right triangle. Draw *BD* perpendicular to *AC* forming right triangles *ABD* and *BDC* (with right angle at *D*). Show that

$$\frac{\sin A}{a} = \frac{\sin B}{b} = \frac{\sin C}{c}.$$

57.   Show that the area of a right triangle is given by $\frac{1}{2}bc \sin A$.

58.   A room measures $9' \times 12' \times 8'$. Compute the angle that the diagonal makes with each of the edges.

4.3 APPLICATIONS OF RIGHT-TRIANGLE TRIGONOMETRY *(We Can Show You a Short Cut, Mister)*

The solution of right triangles is necessary in a variety of situations. The first one we'll consider has to do with an observer looking at an object. The *angle of depression* is the acute angle measured down from the horizontal line to the line of sight, as shown.

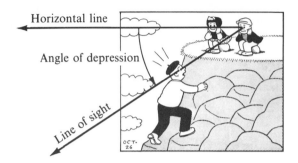

On the other hand, if we take the mountain climber's viewpoint and measure from the horizontal up to the line of sight, we call the angle the *angle of elevation.*

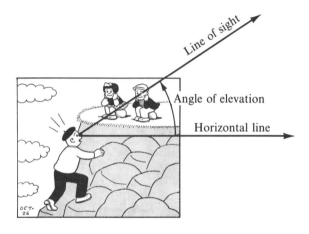

EXAMPLE: The angle of elevation of a tree from a point on the ground 42 m from its base is 33°. Find the height of the tree.

Solution: Let θ = angle of elevation;

h = height of tree.

Then,

$$\tan \theta = \frac{h}{42};$$

$$h = 42 \tan 33°$$

$$= 42(0.6494)$$

$$= 27.28.$$

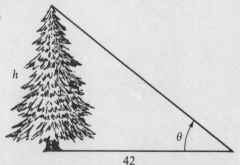

The tree is 27 m tall. (Remember that we are working with two significant digits.)

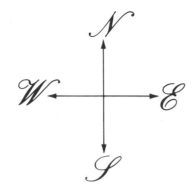

A second application of the solution of right triangles involves the *bearing* of a line, which is defined as an acute angle made with a north-south line. When giving the bearing of a line, we first write N or S to determine whether we measure the angle from the north or the south side of a point on the line. Then we give the measure of the angle followed by E or W, denoting which side of the north-south line we are measuring. Some examples are shown in Figure 4.2.

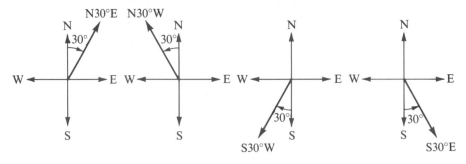

FIGURE 4.2

EXAMPLE: To find the width AB of a canyon, a surveyor measures 100 m from A in the direction of N42.6°W to locate point C. The surveyor then determines that the bearing of CB is N73.5°E. Find the width of the canyon if the point B is situated so that $\angle BAC = 90.0°$.

Solution: Let $\theta = \angle BCA$ in Figure 4.3.

Then $\tan \theta = AB/AC$. So

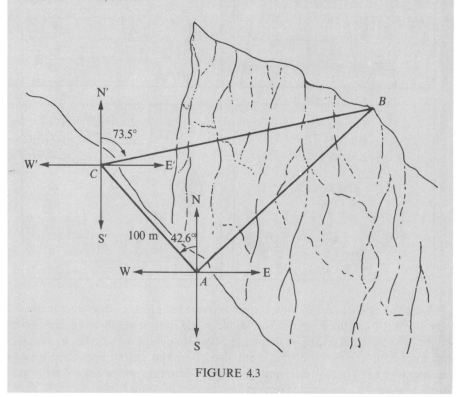

FIGURE 4.3

$$AB = AC \tan \theta$$
$$= 100 \tan \theta.$$

$\angle BCE' = 16.5°$ Complementary angles

$\angle ACS' = 42.6°$ Alternate interior angles

$\angle E'CA = 47.4°$ Complementary angles

$\theta = \angle BCA = \angle BCE' + \angle E'CA$
$$= 16.5° + 47.4°$$
$$= 63.9°$$

Thus,
$$AB = 100 \tan 63.9°$$
$$= 204.125.$$

The canyon is 204 m across (to three significant digits).

PROBLEM SET 4.3

A Problems

1. The angle of elevation of a building from a point on the ground 30 m from its base is 38°. Find the height of the building.

2. The angle of elevation of the top of the Great Pyramid of Khufu (or Cheops) from a point on the ground 351 ft from a point directly below the top is 52.0°. Find the height of the pyramid.

3. From a cliff 150 m above the shoreline, the angle of depression of a ship is 37°. Find the distance of the ship from a point directly below the observer.

4. From a police helicopter flying at 1000 ft, a stolen car is sighted at an angle of depression of 71°. Find the distance of the car from a point directly below the helicopter.

5. To find the east-west boundary of a piece of land, a surveyor must divert his path from point C on the boundary by proceeding due south for 300 ft to a point A. Point B, which is due east of point C, is now found to be in the direction of N49°E from point A. What is the distance CB?

6. To find the distance across a river that runs east-west, a surveyor locates points P and Q on a north-south line on opposite sides of the river. She then paces out 150 ft from Q due east to a point R. Next she determines that the bearing of RP is N58°W. How far is it across the river?

7. A 16-ft ladder on level ground is leaning against a house. If the angle of elevation of the ladder is 52°, how far above the ground is the top of the ladder?

8. How far is the ladder in Problem 7 from the house?

9. If the ladder in Problem 7 is moved so that the bottom is 9 ft from the house, what will be the angle of elevation?

10. Find the height of the Barrington Space Needle if the angle of elevation at 1000 ft from a point on the ground directly below the top is 58.15°.

11. The world's tallest chimney is the 5.5-million-dollar stack of the International Nickel Company. Find the height if the angle of elevation at 1000 ft from a point on the ground directly below the top of the stack is 51.36°.

B Problems

12. a. The angle of elevation of a radio tower from a point on the ground 2.0 km from its base is 1.43°. How tall is the tower in meters?
    b. The angle of elevation of a radio tower from a point on the ground 2.0 mi from its base is 0.81°. How tall is the tower in feet?
    c. Do you think part (a) in meters or part (b) in feet was easier to work? Why?

13. To find the boundary of a piece of land, a surveyor must divert his path from a point *A* on the boundary for 500 ft in the direction S50°E. He then determines that the bearing of a point *B* located directly south of *A* is S40°W. Find the distance *AB*.

14. To find the distance across a river, a surveyor locates points P and Q on either side of the river. Next she measures 100 m from point Q in the direction of S35°E to point R. Then she determines that point P is now in the direction of N25.0°E from point R and that angle PQR is a right angle. Find the distance across the river.

15. If the Empire State Building and the Sears Tower were situated 1000 ft apart, the angle of depression from the top of the Sears Tower to the top of the Empire State Building would be 11.53° and the angle of depression to the foot of the Empire State Building would be 55.48°. Find the heights of the buildings.

16. On the top of the Empire State Building is a TV tower. From a point 1000 ft from a point on the ground directly below the top of the tower the angle of elevation to the bottom of the tower is 51.34° and to the top of the tower is 55.81°. What is the length of the TV tower?

17. Two horseback riders wish to calculate the distance to a very, very exclusive restaurant (called point R). Their present location is point P. They determine that, if they travel S65.4°W for 250 ft to a new location N, the bearing of NR will be N53.7°E. How far is the restaurant from their present location if angle RPN is a right angle?

"Very, very exclusive."

18. A wheel 5.00 ft in diameter rolls up a 15.0° incline. What is the height of the center of the wheel above the base of the incline after the wheel has completed one revolution?

19. What is the height of the center of the wheel in Problem 18 after three revolutions?

20. If the ridge in Figure 4.4 is to be 5.0 ft above point R, how far should the plate be placed from the R directly below the ridge so that the common rafter will have an angle of elevation of 14°? What will the pitch of the common rafter be if it forms this angle? (The pitch is the slope of the roof.)

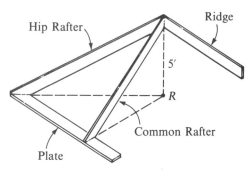

FIGURE 4.4

21. Answer the questions posed in Problem 20 for an angle of elevation of 22.6°.

22. If the distance from the earth to the sun is 92.9 million miles and the angle formed between Venus, the earth, and the sun (as shown in Figure 4.5) is 47.0°, find the distance from the sun to Venus. (See also Problem 31.)

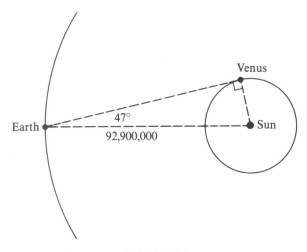

FIGURE 4.5

23. Use the information in Problem 22 to find the distance from the earth to Venus.

24. The largest ground area covered by any office building is that of the Pentagon in Arlington, Virginia. If the radius of the circumscribed circle is 783.5 ft, find the length of one side of the Pentagon.

25. Use the information in Problem 24 to find the radius of the circle inscribed in the Pentagon.

26. To determine the height of the building shown in Figure 4.6, we select a point P and find that the angle of elevation is 59.64°. We then move out a distance of 325.4 ft (on a level plane) to point Q and find that the angle of elevation is now 41.32°. Find the height h of the building.

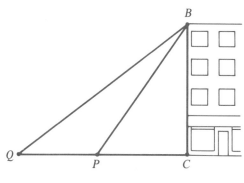

FIGURE 4.6

27. Using Figure 4.6, let the angle of elevation at P be α and at Q be β, and let the distance from P to Q be d. If h is the height of the building, show that

$$h = \frac{d \sin \alpha \sin \beta}{\sin(\alpha - \beta)}.$$

28. A 6.0-ft person is casting a shadow of 4.2 ft. What time is it if the sun rose at 6:15 and is directly overhead at 12:15?

29. How long will the shadow of the person in Problem 28 be at 8:00 A.M.?

30. From the top of a tower 100 ft high, the angles of depression to two landmarks on the plane upon which the tower stands are 18.5° and 28.4°.
    a. Find the distance between the landmarks when they are on the same side of the tower.
    b. Find the distance between the landmarks when they are on opposite sides of the tower.

Mind Bogglers

31. Consult the article "Mathematical Astronomy," by Vincent J. Motto, in the February 1975 issue of the *Two Year College Mathematics Journal*, and then calculate the distance from the earth to Mars, Jupiter, and Saturn.

32. Show that in every right triangle the value of c lies between $(a + b)/\sqrt{2}$ and $a + b$.

4.4 VECTORS *(On to Vectory!)*

Many applications of trigonometry involve quantities that have *both* magnitude and direction, such as forces, velocities, accelerations, and displacements. For example, consider the following.

	Magnitude	Direction
1. An airplane travels with	an air speed of 790 mph	230°.*
2. A car travels at	42 mph	due west.
3. The wind blows at	46.1 mph	S63°W.
4. An object on an inclined plane weighs	450 lb	down (gravity).
5. A man is pulling a rope (force)	130 lb	42° from the vertical.

In mathematics we represent such quantities as *vectors*.

A vector is a line segment that has both magnitude and direction.

The length of the vector represents the *magnitude* of the quantity being represented, and the *direction* of the vector represents the direction of the quantity. Two vectors are equal if they have the same magnitude and direction.

Let's choose a point O and call it the *origin*. For convenience we'll consider vectors as directed line segments from point O to some point $P(x, y)$ in the plane. This vector is denoted by $\overrightarrow{OP}$ or **v**. (In the text we use boldface. In your work you will write $\vec{v}$.) The *magnitude* of $\overrightarrow{OP}$ is denoted by $|\overrightarrow{OP}|$ or $|\mathbf{v}|$. The zero vector, denoted by **0**, is a vector whose magnitude is 0.

If **v** and **w** represent any two vectors having different directions, then the *sum* or *resultant* is the vector drawn as the diagonal of the parallelogram having **v** and **w** as the adjacent sides, as shown in Figure 4.7. The vectors **v** and **w** are called *components*.

In physics it can be shown that two forces acting simultaneously at the same point are combined according to this definition of addition.

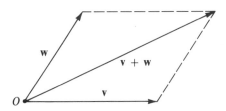

FIGURE 4.7   Resultant of vectors **v** and **w**.

*In aviation the bearing of a point is measured in a clockwise direction from the north.

If two vectors form a right angle, they are called *rectangular components*, and, given any nonzero vector $\overrightarrow{OP}$, it is possible to find (or *resolve*) the vector into the sum of two vectors in two specified directions. In this chapter we'll work with rectangular components, and in the next chapter we'll generalize to include any two vectors. For example, if a given vector has magnitude 5.00 and its direction is given by $\theta = 30.0°$, where θ is the angle the vector makes with the x-axis, we can resolve this vector into two components.

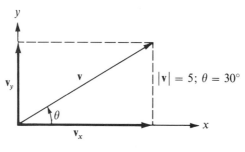

FIGURE 4.8

Let $\mathbf{v}_x$ be the horizontal component and $\mathbf{v}_y$ the vertical component. Since a right triangle is formed, we see the following.

$$\cos \theta = \frac{|\mathbf{v}_x|}{|\mathbf{v}|} \qquad\qquad \sin \theta = \frac{|\mathbf{v}_y|}{|\mathbf{v}|}$$

$$|\mathbf{v}_x| = |\mathbf{v}| \cos \theta \qquad\qquad \mathbf{v}_y = |\mathbf{v}| \sin \theta$$

$$= 5 \cos 30° \qquad\qquad = 5 \sin 30°$$

$$= \frac{5}{2}\sqrt{3} \qquad\qquad = \frac{5}{2}$$

$$= 4.33 \qquad\qquad = 2.50 \qquad \text{(to three significant digits)}$$

Thus, $\mathbf{v}_x$ is a horizontal vector with magnitude 4.33 and $\mathbf{v}_y$ is a vertical vector with magnitude 2.50. For ease in naming vectors, we sometimes denote a vector $\overrightarrow{OP}$ by (x, y), where (x, y) are the coordinates of P. Thus,

$$\mathbf{v}_x = (4.33, 0) \quad \text{and} \quad \mathbf{v}_y = (0, 2.50).$$

EXAMPLE: A swimmer sets out on an easterly course and swims at a constant 3.5 mph. The current is carrying him due south at 7.2 mph. What is the true bearing of his course, and how fast is he actually traveling?

Solution:

Step 1: Draw a vector $\mathbf{v}$ representing the swimmer's course. The direction is east, and the magnitude is 3.5.

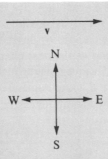

Step 2: Draw a vector **w** representing the current (direction south; magnitude 7.2).

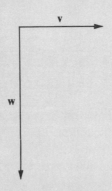

Step 3: The solution is the vector **v** + **w** (called the resultant).

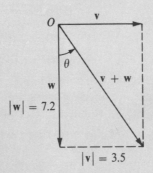

The direction is obtained by finding θ.

$$\tan \theta = \frac{|\mathbf{v}|}{|\mathbf{w}|}$$

$$= \frac{3.5}{7.2}$$

$$\theta = \tan^{-1}\left(\frac{3.5}{7.2}\right)$$

$$= 25.92°$$

$$= 26° \quad \text{(to two significant digits)}$$

The magnitude $|v + w| = \sqrt{|v|^2 + |w|^2}$

$$= \sqrt{3.5^2 + 7.2^2}$$

$$= \sqrt{12.25 + 51.84}$$

$$= \sqrt{64.09}$$

$$= 8.0. \qquad \text{(to two significant digits)}$$

The swimmer is traveling S26°E at 8 mph.

In general, if we are given two mutually perpendicular vectors v and w and the resultant vector $v + w$ forming an angle θ with vector v, we can let $|v| = a$, $|w| = b$, and $|v + w| = r$.

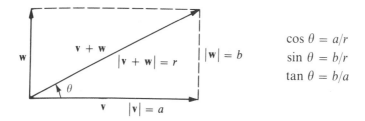

$$\cos \theta = a/r$$
$$\sin \theta = b/r$$
$$\tan \theta = b/a$$

Then, to find the resultant:

$$\theta = \tan^{-1}\left(\frac{b}{a}\right);$$

$$r = \sqrt{a^2 + b^2}.$$

To resolve the vector $v + w$:

$$a = r \cos \theta;$$

$$b = r \sin \theta.$$

EXAMPLE: What is the smallest force necessary to keep a 3420-lb car from rolling down a hill that makes an angle of 5.25° with the horizontal? What is the force of the car against the hill? (In this book we'll assume that friction is negligible in problems like this.)

Solution: The weight (or force) of the car, $\overrightarrow{OC}$, acts vertically downward. This vector can be resolved into two vectors: $\overrightarrow{OH}$, the force of the car against the hill, and $\overrightarrow{OD}$, the force pushing the car down the hill. (See Figure 4.9.)

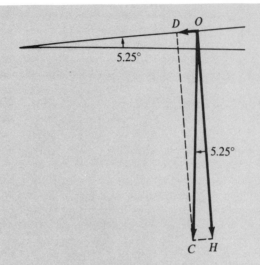

FIGURE 4.9

Notice that $\vec{OC}$ and $\vec{OH}$ are perpendicular to the sides making up the 5.25° angle; thus, $\angle COH = 5.25°$. Therefore, we have the following.

$$\sin 5.25° = \frac{|\vec{OD}|}{|\vec{OC}|} \qquad \cos 5.25° = \frac{|\vec{OH}|}{|\vec{OC}|}$$

$$|\vec{OD}| = |\vec{OC}| \sin 5.25° \qquad |\vec{OH}| = |\vec{OC}| \cos 5.25°$$
$$= 312.9 \qquad\qquad = 3406$$

Thus a force of 312.9 lb is necessary to keep the car from rolling down the hill, and the car is pushing against the hill with a force of 3406 lb.

PROBLEM SET 4.4

A Problems

In Problems 1–6 use the vectors shown in Figure 4.10 to fill in the blanks (to two significant digits).

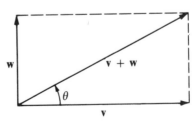

FIGURE 4.10

	$|\mathbf{v}|$	$|\mathbf{w}|$	$|\mathbf{v} + \mathbf{w}|$	θ
1.	3.0	4.0		
2.	4.0	3.0		
3.	14.0	18.0		
4.			10	30°
5.			23	18°
6.			120	27°

7. A woman sets out in a rowboat on a due-west course and rows at 4.8 mph. The current is carrying the boat due south at 12 mph. What is the true bearing of the rowboat, and how fast is the boat actually traveling?

8. An airplane is headed due west at 240 mph. The wind is blowing due south at 43 mph. What is the true bearing of the plane, and how fast is it actually traveling? Note: In aviation the bearing of a point is measured in a clockwise direction from the north.

9. What is the minimum force necessary to keep a boxcar weighing 52.0 tons from rolling down the world's steepest standard-gauge grade, which measures 5.2° (located between the Samala River Bridge and Zunil in Guatemala)? What is the force of the boxcar against the hill?

10. What is the minimum force necessary to keep a 250-lb barrel from rolling down an inclined plane making an angle of 12° with the horizontal? What is the force of the barrel against the plane?

B Problems

11. An airplane is heading 215° with a velocity of 723 mph. How far south has it traveled in one hour? (See note in Problem 8.)

12. An airplane is heading 43.0° with a velocity of 248 mph. How far east has it traveled in two hours? (See note in Problem 8.)

13. A cannon is fired at an angle of 10.0° above the horizontal, with the initial speed of the cannonball 2120 fps. Find the magnitude of the vertical and horizontal components of the initial velocity.

14. Answer Problem 13 for an angle of 17.8°.

15. A force of 253 pounds is necessary to keep a weight of exactly 400 pounds from sliding down an inclined plane. What is the angle of inclination of the plane?

16. A force of 486 pounds is necessary to keep a weight of exactly 800 pounds from sliding down an inclined plane. What is the angle of inclination of the plane?

17. A pilot is flying at an air speed of 241 mph in a wind blowing 20.4 mph from the east. In what direction must the pilot head in order to fly due north? What is the pilot's speed relative to the ground? (See note in Problem 8.)

18. Answer the questions posed in Problem 17 with the pilot wishing to fly due south. (See note in Problem 8.)

19. A particle has an acceleration of 4.32 ft/sec^2 in the direction making an angle of 42.5° with the x-axis. Find the horizontal and vertical components of its acceleration.

20. A particle has an acceleration of 2.41 ft/sec^2 in the direction making an angle of 38.4° with the x-axis. Find the horizontal and vertical components of its acceleration.

21. A cable that can withstand 5250 pounds is used to pull cargo up an inclined ramp for storage. If the inclination of the ramp is 25.5°, find the heaviest piece of cargo that can be pulled up the ramp.

22. Answer the question posed in Problem 21 for a ramp whose angle of inclination is 18.2°.

23. Suppose an object is thrown from a car traveling due east at 51 km per hour. If the object is thown due south at 200 m per minute, indicate the direction and speed of the object.

24. Work Problem 23 for a car traveling at 51 mph and an object thrown at 200 ft/min.

Mind Bogglers

25. A 50.0-kg weight is hanging from a cable, which is held away from the vertical by a 2.00-m foot brace 1.00 m below the point where the cable is attached to the vertical.
    a. Find the force of the cable between the vertical and the end of the brace.
    b. Find the force in the brace.

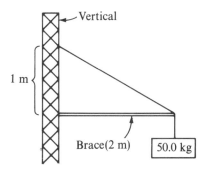

26. Consider three forces: 214.3 kg in the direction S18.46°W, 148.9 kg in the direction S71.54°E, and 198.7 kg in the direction S73.67°W. Suppose we wish to find a force acting due north and another force acting due east that will counterbalance the three given forces. What are these two forces?

4.5 SUMMARY AND REVIEW

TERMS

Angle of depression (4.3)   Bearing of a line (4.3)
Angle of elevation (4.3)   Components (4.4)

Direction of a vector (4.4)   Resultant vector (4.4)
Magnitude of a vector (4.4)   Significant digits (4.2)
Rectangular components (4.4)   Solving a triangle (4.1)
Resolving vectors (4.4)   Vector (4.4)

CONCEPTS

(4.1)   If θ is an acute angle in a right triangle,

$$\sin \theta = \frac{\text{opposite side}}{\text{hypotenuse}} ; \quad \cos \theta = \frac{\text{adjacent side}}{\text{hypotenuse}} ; \quad \tan \theta = \frac{\text{opposite side}}{\text{adjacent side}}.$$

(4.1)   Solving a right triangle.

(4.4)   Let $\mathbf{v}$ and $\mathbf{w}$ be any vectors with θ the angle between $\mathbf{v} + \mathbf{w}$ and $\mathbf{v}$. Also, let $|\mathbf{v}| = a$, $|\mathbf{w}| = b$, and $|\mathbf{v} + \mathbf{w}| = r$.

Given $\mathbf{v}$ and $\mathbf{w}$, find $|\mathbf{v} + \mathbf{w}|$:   Given $\mathbf{v} + \mathbf{w}$, find $|\mathbf{v}|$ and $|\mathbf{w}|$:

$$\theta = \tan^{-1}(b/a);$$ $$a = r \cos \theta;$$

$$r = \sqrt{a^2 + b^2}$$ $$b = r \sin \theta.$$

PROBLEM SET 4.5 *(Chapter 4 Test)*

(4.1) *Complete the following table. Assume that $\triangle ABC$ has a right angle at C and that the hypotenuse is c. Some answers have been provided as examples.*

		To Find	Known		Procedure
1.		a	b	c	$\sqrt{c^2 - b^2}$
	a.	a	A	b	
	b.	a	A	c	
	c.	a	B	b	
	d.	a	B	c	
2.	a.	b	a	c	
	b.	b	A	a	
		b	A	c	$c \cos A$
	c.	b	B	a	
	d.	b	B	c	

(table continued)

		To Find	Known	Procedure
3.	a.	c	a b	
	b.	c	A a	
	c.	c	A b	
	d.	c	B a	
		c	B b	$b/\sin B$ or $b \csc B$
4.		A	a b	$\tan^{-1}(a/b)$
	a.	A	a c	
	b.	A	b c	
		A	B a	$90° - B$
	c.	A	B b	
	d.	A	B c	
5.	a.	B	a b	
	b.	B	a c	
	c.	B	b c	
	d.	B	A a	
		B	A b	$90° - A$
		B	A c	$90° - A$

6. (4.2) Use a calculator or logarithms to simplify the given expressions. Round off your answers to the appropriate number of digits.

   a. $\sqrt{2.48^2 + 4.62^2}$

   b. $\dfrac{(5.21)^2 + 3.481}{6.2}$

   c. $\dfrac{241 \sin 61.5°}{1125}$

   d. $\sin^{-1}\left(\dfrac{23.0 \sin 48.3°}{27.4}\right)$

7. (4.3) The tallest human-built structure in the world is the TV transmitter tower of KTHI-TV in North Dakota. Find its height if the angle of elevation at 501.0 ft is 76.35°.

8. (4.3) To measure the distance across the Rainbow Bridge in Utah, a surveyor selected two points P and Q on either end of the bridge. From point Q the surveyor measured 500 ft in the direction N38.4°E to point R. Point P was then determined to be in the

direction S67.5°W. What is the distance across the Rainbow Bridge if all the preceding measurements are in the same plane and angle *PQR* is a right angle?

9. (4.3) When viewing Angel Falls (the world's highest waterfall) from Observation Platform *A*, located on the same level as the bottom of the falls, we calculate the angle of elevation to the top of the falls to be 69.30°. From Observation Platform *B*, which is located on the same level exactly 1000 ft from the first observation point, we calculate the angle of elevation to the top of the falls to be 52.90°. How high are the falls?

10. (4.4) An expendable object is hurled from a catapult due east with a velocity of 38 feet per second (fps). If the wind is blowing due south at 15 mph (= 22 fps), what are the true bearing and the velocity of the expendable object?

In the last chapter we studied the solution of right triangles. In this chapter we turn to the solution of *oblique* triangles, or triangles that are not right triangles. We will derive two trigonometric laws, called the *law of cosines* and the *law of sines,* which will allow us to solve certain triangles, provided that we can use a calculator or have access to trigonometric tables and logarithms. If no calculator is available, there is a third trigonometric law that lends itself to logarithmic calculation. It is called the *law of tangents.*

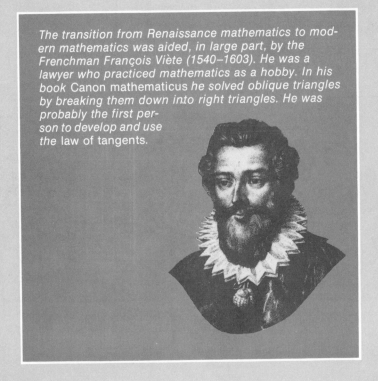

The transition from Renaissance mathematics to modern mathematics was aided, in large part, by the Frenchman François Viète (1540–1603). He was a lawyer who practiced mathematics as a hobby. In his book Canon mathematicus *he solved oblique triangles by breaking them down into right triangles. He was probably the first person to develop and use the* law of tangents.

5

APPLICATIONS OF TRIGONOMETRY

Oblique Triangles

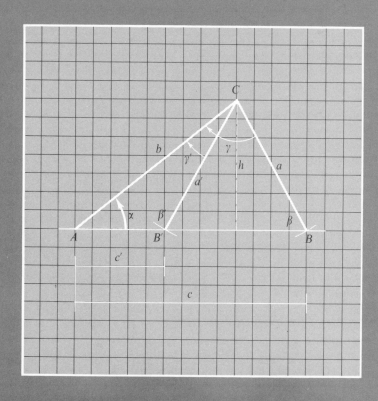

5.1 OBLIQUE TRIANGLES *(If It's Not Right, Then It's Oblique)*

We will approach the solution of *oblique triangles* (triangles with no right angles) by studying the possible combinations of given information. As in the previous chapter, we'll consider a triangle labeled as in Figure 5.1. However, to avoid confusion, we'll agree that angle A is also called α, angle B is also called β, and angle C is also called γ.

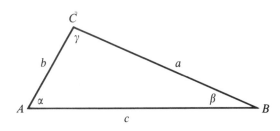

FIGURE 5.1

In general, we'll be given three parts of the triangle and be asked to find the remaining three parts. But can we do so given *any* three parts? Let's list the possibilities.

1. SSS   By SSS we mean that we're given three sides and want to find the three angles.
2. SAS   We are given two sides and an included angle.
3. ASA or AAS   We are given two angles and a side.
4. SSA   We are given an angle and a side opposite the angle, as well as another side.
5. AAA   We are given three angles.

We'll consider these possibilities one at a time. For SSS it is necessary for the sum of the lengths of the two smaller sides to be greater than the length of the largest side. If this is the case, we use a generalization of the Pythagorean Theorem called the *law of cosines*:

In triangle ABC labeled as shown in Figure 5.1,

$$c^2 = a^2 + b^2 - 2ab \cos \gamma.$$

Let angle C be an angle in standard position with A on the positive x-axis, as shown in Figure 5.2. The coordinates of the vertices are as follows.

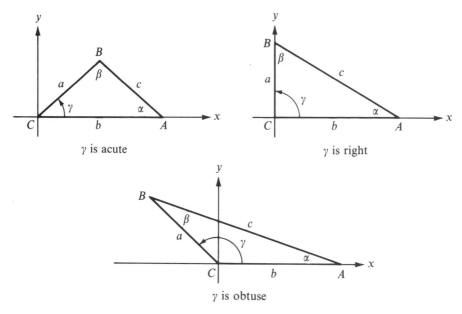

γ is acute γ is right

γ is obtuse

FIGURE 5.2

$C(0, 0)$	Since C is in standard position.
$A(b, 0)$	Since A is on the x-axis a distance of b units from the origin.
$B(a \cos \gamma, a \sin \gamma)$	Draw a perpendicular from B to the x-axis, and let $B(x, y)$. By definition of the trigonometric functions, $\cos \gamma = x/a$ and $\sin \gamma = y/a$. Thus, $x = a \cos \gamma$ and $y = a \sin \gamma$.

We now use the distance formula for the distance between $A(b, 0)$ and $B(a \cos \gamma, a \sin \gamma)$.

$$\begin{aligned} c^2 &= (a \cos \gamma - b)^2 + (a \sin \gamma - 0)^2 \\ &= a^2 \cos^2 \gamma - 2ab \cos \gamma + b^2 + a^2 \sin^2 \gamma \\ &= a^2(\cos^2 \gamma + \sin^2 \gamma) + b^2 - 2ab \cos \gamma \\ &= a^2 + b^2 - 2ab \cos \gamma \end{aligned}$$

Notice that for a right triangle $\gamma = 90°$, which means that

$$c^2 = a^2 + b^2 - 2ab \cos 90°$$

or

$$c^2 = a^2 + b^2,$$

since $\cos 90° = 0$. But this last equation is simply the Pythagorean Theorem; thus the law of cosines is a generalization of the familiar Pythagorean Theorem.

By letting A and B, respectively, be in standard position, we can show the following.

$$a^2 = b^2 + c^2 - 2bc \cos \alpha$$
$$b^2 = a^2 + c^2 - 2ac \cos \beta$$

We can now solve for α, β, or γ to find the angles when we're given three sides:

Law of Cosines

$$a^2 = b^2 + c^2 - 2bc \cos \alpha \qquad \cos \alpha = \frac{b^2 + c^2 - a^2}{2bc}$$

$$b^2 = a^2 + c^2 - 2ac \cos \beta \qquad \cos \beta = \frac{a^2 + c^2 - b^2}{2ac}$$

$$c^2 = a^2 + b^2 - 2ab \cos \gamma \qquad \cos \gamma = \frac{a^2 + b^2 - c^2}{2ab}$$

EXAMPLE: What is the smallest angle of a triangular patio whose sides measure 25, 18, and 21 feet?

Solution: If γ represents the smallest angle, then c (the side opposite γ) must be the smallest side, so $c = 18$. Then:

$$\cos \gamma = \frac{a^2 + b^2 - c^2}{2ab}$$

$$= \frac{25^2 + 21^2 - 18^2}{2(25)(21)}; \qquad \longleftarrow \text{ Use this number and trig tables if you have only a four-function calculator.}$$

$$\gamma = \cos^{-1}\left(\frac{25^2 + 21^2 - 18^2}{2(25)(21)}\right).$$

If you're using a calculator with algebraic logic, you'd have the following.

| 25 | × | 25 | + | 21 | × | 21 | − | 18 | × | 18 | = | ÷ | 2 | ÷ | 25 | ÷ |

or [25] [x²]   or [21] [x²]   or [18] [x²]

| 21 | = | INV | COS |

The result shown is 45.03565072. The answer correct to two significant digits is 45°. (If your calculator has an x^2 key, be sure to use it.)

If you're using a calculator with reverse polish logic, you'd have the following.

| 25 | ENTER | 25 | × | 21 | ENTER | 21 | × | + | 18 | ENTER | 18 |

| × | − | 2 | ENTER | 25 | × | 21 | × | ÷ | ARC | COS |

The result shown is 45.0356071. (Notice that small differences may result on different calculators.) To two significant digits, the answer is 45°.

If you do not have a calculator available, the law of cosines will not be suitable for logarithmic calculations. However, notice that:

$$\cos \gamma = \frac{a^2 + b^2 - c^2}{2ab};$$

$$\cos \gamma + 1 = \frac{a^2 + b^2 - c^2}{2ab} + 1$$

$$= \frac{a^2 + b^2 - c^2 + 2ab}{2ab}$$

$$= \frac{(a + b - c)(a + b + c)}{2ab}.$$

Thus

$$\cos \gamma + 1 = \frac{(25 + 21 - 18)(25 + 21 + 18)}{2(25)(21)}$$

can be solved using logarithms.

The next example illustrates a special computational problem that you may encounter on certain problems when using Table III.

EXAMPLE: Find α where $a = 7.0$, $b = 4.0$, and $c = 5.0$.

Solution: By the law of cosines:

$$\cos \alpha = \frac{b^2 + c^2 - a^2}{2bc}$$

$$= \frac{16 + 25 - 49}{2(4)(5)}$$

$$= \frac{-8}{2(4)(5)}$$

$$= -0.2.$$

By calculator:

$$\alpha = \cos^{-1}(-0.2)$$

$$= 101.536959.$$

However, by table we must first find a reference angle for α: 78.5°. We find this angle by using 0.2 instead of -0.2, since Table III does not give negative values. Then, because of the negative value of the cosine, we find $180° - 78.5° = 101.5°$. Thus, $\alpha = 102°$.

The second possibility listed for solving oblique triangles is that of being given two sides and an included angle. It is necessary that the given angle be less than

180°. We again use the law of cosines for this possibility, as shown by the following example.

EXAMPLE: Find c where $a = 52.0$, $b = 28.3$, and $\gamma = 28.5°$.

Solution: By the law of cosines:

$$c^2 = a^2 + b^2 - 2ab \cos \gamma$$
$$= (52.0)^2 + (28.3)^2 - 2(52.0)(28.3) \cos 28.5°.$$

By calculator:

$$c^2 = 918.355474;$$
$$c = 30.30438044.$$

To three significant digits, $c = 30.3$.

If you do not have a calculator available, you should use the law of tangents on this problem, since the law of cosines does not lend itself to logarithmic calculations. The law of tangents is discussed in Problems 15–17 on page 186.

The other possibilities listed for solving oblique triangles use the law of sines, which will be discussed in the next section.

PROBLEM SET 5.1

A Problems

Solve triangle ABC for the parts requested in Problems 1–10.

1. $a = 7.0$; $b = 8.0$; $c = 2.0$. Find α.

2. $a = 7.0$; $b = 5.0$; $c = 4.0$. Find β.

3. $a = 10$; $b = 4.0$; $c = 8.0$. Find γ.

4. $a = 4.00$; $b = 5.00$; $c = 6.00$. Find α.

5. $a = 11$; $b = 9.0$; $c = 8.0$. Find the largest angle.

6. $a = 18$; $b = 25$; $\gamma = 30°$. Find c.

7. $a = 15$; $b = 8.0$; $\gamma = 38°$. Find c.

8. $a = 18$; $c = 11$; $\beta = 63°$. Find b.

9. $b = 14$; $c = 12$; $\alpha = 82°$. Find a.

10. $b = 21$; $c = 35$; $\alpha = 125°$. Find a.

B Problems

11. Prove that $a^2 = b^2 + c^2 - 2bc \cos \alpha$.

12. Prove that $b^2 = a^2 + c^2 - 2ac \cos \beta$.

13. From the equation given in Problem 11 show that

$$\cos \alpha + 1 = \frac{(b + c - a)(a + b + c)}{2bc}.$$

14. From the equation given in Problem 12 show that

$$\cos \beta + 1 = \frac{(a + c - b)(a + b + c)}{2ac}$$

Solve triangle ABC for the parts requested in Problems 15–20. If the triangle cannot be solved, tell why.

15. $a = 38$; $b = 41$; $c = 25$. Find the largest angle.

16. $a = 45$; $b = 92$; $c = 41$. Find the smallest angle.

17. $a = 241$; $b = 187$; $c = 100$. Find β.

18. $a = 38.2$; $b = 14.8$; $\gamma = 48.2°$. Find c.

19. $b = 123$; $c = 485$; $\alpha = 163.0°$. Find a.

20. $a = 48.3$; $c = 35.1$; $\beta = 215.0°$. Find b.

5.2 LAW OF SINES *(Don't Break the Law—of Sines)*

In the last section we listed five possibilities for given information in solving oblique triangles (see 1–5, page 178) and discussed the first two cases. In this section we'll consider cases 3 and 5.

Case 5 supposes that three angles are given. However, from what we know of similar triangles (see Figure 5.3), we can conclude that we cannot solve the triangle without knowing the length of at least one side.

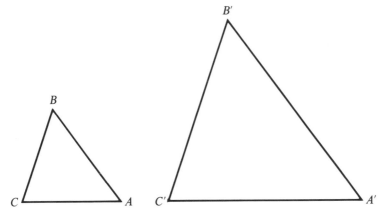

FIGURE 5.3   Similar Triangles. Similar triangles have the same shape but not necessarily the same size. That is, corresponding angles of similar triangles have equal measure.

Case 3 supposes that two angles and a side are given. For a triangle to be formed, the sum of the two given angles must be less than 180°, and the given side must be greater than zero. If we know two angles, we can easily find the third angle, since the sum of the three angles is 180°. The law of cosines is not sufficient in this case because we need at least two sides.

Let's consider any oblique triangle, as shown in Figure 5.4.

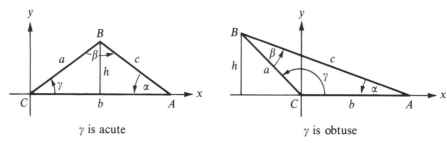

γ is acute γ is obtuse

FIGURE 5.4

We'll let h = the height of the triangle with base CA. Then

$$\sin \alpha = \frac{h}{c} \quad \text{and} \quad \sin \gamma = \frac{h}{a}.$$

Solving for h:

$$h = c \sin \alpha \quad \text{and} \quad h = a \sin \gamma.$$

Thus,

$$c \sin \alpha = a \sin \gamma.$$

Dividing by ac:

$$\frac{\sin \alpha}{a} = \frac{\sin \gamma}{c}.$$

If we repeat these steps for the height of the same triangle with base AB (see Problem 10 of Problem Set 5.2),

$$\frac{\sin \alpha}{a} = \frac{\sin \beta}{b}.$$

This result is called the *law of sines*.

Law of Sines

In any triangle ABC,

$$\frac{\sin \alpha}{a} = \frac{\sin \beta}{b} = \frac{\sin \gamma}{c}.$$

The above "equation" is written using two equals signs, but any pair of terms will work:

$$\frac{\sin \alpha}{a} = \frac{\sin \beta}{b}; \qquad \frac{\sin \alpha}{a} = \frac{\sin \gamma}{c}; \qquad \frac{\sin \beta}{b} = \frac{\sin \gamma}{c}.$$

EXAMPLE: Solve the triangle in which $a = 20$, $\alpha = 38°$, and $\beta = 121°$.

Solution:

$\alpha = 38°$ Given

$\beta = 121°$ Given

$\gamma = 21°$ Since $\alpha + \beta + \gamma = 180°$,

$\qquad \gamma = 180° - 38° - 121°$

$\qquad = 21°$.

$a = 20$ Given

$b = 28$ Use the law of sines:

$\qquad \dfrac{\sin 38°}{20} = \dfrac{\sin 121°}{b};$

$\qquad b = \dfrac{20 \sin 121°}{\sin 38°}$

$\qquad = \dfrac{20(0.8572)}{0.6157}$ Use tables or a calculator.

$\qquad = 27.85$. Use logarithms or a calculator.

$c = 12$ Use the law of sines:

$\qquad \dfrac{\sin 38°}{20} = \dfrac{\sin 21°}{c};$

$\qquad c = \dfrac{20 \sin 21°}{\sin 38°}$

$\qquad = 11.64$.

Notice that the answers stated above are given to two significant digits.

PROBLEM SET 5.2

A Problems

Solve triangle ABC with the parts given in Problems 1–9. If the triangle cannot be solved, tell why.

1. $a = 10$; $\alpha = 48°$; $\beta = 62°$
2. $a = 18$; $\alpha = 65°$; $\gamma = 115°$
3. $a = 30$; $\beta = 50°$; $\gamma = 100°$
4. $b = 23$; $\alpha = 25°$; $\beta = 110°$
5. $b = 40$; $\alpha = 50°$; $\gamma = 60°$
6. $b = 90$; $\beta = 85°$; $\gamma = 25°$

7. $c = 53; \alpha = 82°; \beta = 19°$ 8. $c = 43; \alpha = 120°; \gamma = 7°$

9. $c = 115; \beta = 81.0°; \gamma = 64.0°$

B Problems

10. Given $\triangle ABC$, show that $(\sin \alpha)/a = (\sin \beta)/b$.

11. Given $\triangle ABC$, show that $(\sin \beta)/b = (\sin \gamma)/c$.

12. Given $\triangle ABC$, show that $(\sin \alpha)/\sin \beta = a/b$.

13. Using Problem 12, show that
$$\frac{\sin \alpha - \sin \beta}{\sin \alpha + \sin \beta} = \frac{a - b}{a + b}.$$

14. Using Problem 13 and the formulas for the sum and difference of sines, show that
$$\frac{2 \cos \frac{1}{2}(\alpha + \beta) \sin \frac{1}{2}(\alpha - \beta)}{2 \sin \frac{1}{2}(\alpha + \beta) \cos \frac{1}{2}(\alpha - \beta)} = \frac{a - b}{a + b}.$$

15. In Section 5.1 we mentioned a *law of tangents*, which is useful in logarithmic calculations. Using Problem 14, show that
$$\frac{\tan \frac{1}{2}(\alpha - \beta)}{\tan \frac{1}{2}(\alpha + \beta)} = \frac{a - b}{a + b}.$$

16. Another form of the *law of tangents* (see Problem 15) is
$$\frac{\tan \frac{1}{2}(\beta - \gamma)}{\tan \frac{1}{2}(\beta + \gamma)} = \frac{b - c}{b + c}.$$
Derive this formula.

17. A third form of the *law of tangents* (see Problems 15 and 16) is
$$\frac{\tan \frac{1}{2}(\alpha - \gamma)}{\tan \frac{1}{2}(\alpha + \gamma)} = \frac{a - c}{a + c}.$$
Derive this formula.

Solve $\triangle ABC$ with the parts given in Problems 18–28. If the triangle cannot be solved, tell why.

18. $a = 41.0; \alpha = 45.2°; \beta = 21.5°$ 19. $b = 55.0; c = 92.0; \alpha = 98.0°$

20. $b = 58.3; \alpha = 120°; \gamma = 68.0°$ 21. $c = 123; \alpha = 85.2°; \beta = 38.7°$

22. $a = 26; b = 71; c = 88$ 23. $\alpha = 48°; \beta = 105°; \gamma = 27°$

24. $a = 25.0; \beta = 81.0°; \gamma = 25.0°$ 25. $a = 25.0; b = 45.0; c = 102$

26. $a = 80.6; b = 23.2; \gamma = 89.2°$ 27. $b = 1234; \alpha = 85.26°; \beta = 24.45°$

28. $c = 28.36; \beta = 42.10°; \gamma = 102.30°$

Mind Bogglers

29. *Newton's Formula* involves all six parts of a triangle. It is not useful in solving a triangle, but it is helpful in checking your results after you have done so. Show that

$$\frac{a + b}{c} = \frac{\cos \frac{1}{2}(\alpha - \beta)}{\sin \frac{1}{2}\gamma}.$$

30. *Mollweide's Formula* involves all six parts of a triangle. It is not useful in solving a triangle, but it is helpful in checking your results after you have done so. Show that

$$\frac{a - b}{c} = \frac{\sin \frac{1}{2}(\alpha - \beta)}{\cos \frac{1}{2}\gamma}.$$

5.3 THE AMBIGUOUS CASE *(Try Not to Be Ambiguous)*

The remaining case of solving oblique triangles as given in Section 5.1 is Case 4, in which we're given two sides and an angle that is not an included angle. For convenience we'll assume that we're given sides a and b and angle α. Under what conditions will a triangle be formed? Let's consider each possibility separately.

1. Suppose that $\alpha \geq 90°$. There are two possibilities.
    i. $a \leq b$
    No triangle is formed (see Figure 5.5).

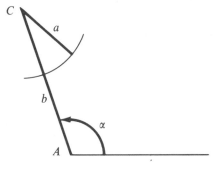

FIGURE 5.5

EXAMPLE: Let $a = 1.0$, $b = 2.0$, and $\alpha = 110°$.

Solution: $\quad \dfrac{\sin \alpha}{a} = \dfrac{\sin \beta}{b}$

$\qquad \dfrac{\sin 110°}{1} = \dfrac{\sin \beta}{2}$

$\qquad\qquad \sin \beta = 2 \sin 110°$

$\qquad\qquad\qquad = 2(0.9397)$

$\qquad\qquad\qquad = 1.8794$

But $\sin \beta > 1$, so there is no solution.

ii. $a > b$

One triangle is formed (see Figure 5.6). Use the law of sines.

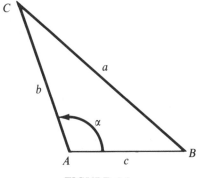

FIGURE 5.6

EXAMPLE: Let $a = 3.0$, $b = 2.0$, and $\alpha = 110°$.

Solution: $\dfrac{\sin \alpha}{a} = \dfrac{\sin \beta}{b}$

$$\frac{\sin 110°}{3} = \frac{\sin \beta}{2}$$

$$\sin \beta = \frac{2}{3} \sin 110°$$

$$= \frac{2}{3}(0.9397)$$

$$= 0.6265$$

$$\beta = 38.79$$

$\alpha = 110°$ Given

$\beta = 39°$ See above.

$\gamma = 31°$ $\gamma = 180° - 110° - 39°$

$a = 3.0$ Given

$b = 2.0$ Given

$c = 1.6$ Use the law of sines:

$$\frac{\sin 110°}{3} = \frac{\sin 31°}{c};$$

$$c = \frac{3 \sin 31°}{\sin 110°}$$

$$= 1.644.$$

2.   Suppose that $\alpha < 90°$. There are four possibilities (letting h be the altitude of $\triangle ABC$ drawn from C to AB).

i. $a < h < b$

No triangle is formed (see Figure 5.7).

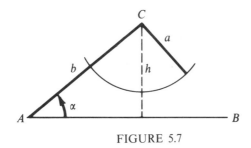

FIGURE 5.7

EXAMPLE: Let $a = 1.0$, $b = 2.0$, and $\alpha = 40°$.

Solution: $\dfrac{\sin \alpha}{a} = \dfrac{\sin \beta}{b}$

$$\dfrac{\sin 40°}{1} = \dfrac{\sin \beta}{2}$$

$$\sin \beta = 2 \sin 40°$$

$$= 2(0.6428)$$

$$= 1.2856$$

But $\sin \beta > 1$, so there is no solution.

ii. $a = h < b$

A right triangle is formed (see Figure 5.8).

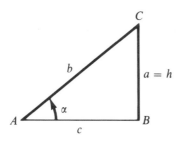

FIGURE 5.8

EXAMPLE: Let $a = b \sin \alpha$, $b = 2.00$, and $\alpha = 40.0°$.

Solution: Since $\sin \alpha = h/b$, we see that, if $a = h$, then $a = b \sin \alpha$.

$\alpha = 40.0°$ Given

$\beta = 90.0°$ Since $a = h$

$\gamma = 50.0°$ Since $\gamma = 90° - \alpha$

$a = 1.29$ $a = 2 \sin 40°$

$\qquad = 1.2856$

$b = 2.00$ Given

$c = 1.53$ $a^2 + c^2 = b^2$ in a right triangle with right angle at B.

$$c = \sqrt{b^2 - a^2}$$

$$= \sqrt{2^2 - 1.29^2}$$

$$= 1.5284$$

iii. $h < a < b$

Two triangles are formed (see Figure 5.9). This situation is called the *ambiguous case* and is really the only special case you must watch for (all the other cases can be determined from the calculations without any special consideration).

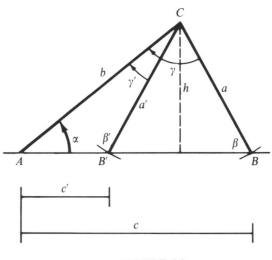

FIGURE 5.9

EXAMPLE: Let $a = 1.50$, $b = 2.00$, and $\alpha = 40.0°$.

Solution: $\dfrac{\sin \alpha}{a} = \dfrac{\sin \beta}{b}$

$$\frac{\sin 40°}{1.5} = \frac{\sin \beta}{2}$$

$$\sin \beta = \frac{2 \sin 40°}{1.5}$$

$$= \frac{4}{3}(0.6428)$$

$$= 0.8570$$

$$\beta = 59.0°$$

But from Figure 5.9 we can see that this is only the acute-angle solution. For the obtuse-angle solution—call it β'—we find

$$\beta' = 180° - \beta$$

$$= 121°.$$

We finish the problem by working two calculations, which are presented side by side.

Solution 1			*Solution 2*	
$\alpha = 40.0°$	Given		$\alpha = 40.0°$	Given
$\beta = 59.0°$	See above.		$\beta' = 121°$	See above.
$\gamma = 81.0°$	$\gamma = 180° - \alpha - \beta$		$\gamma' = 19.0°$	$\gamma' = 180° - \alpha - \beta'$
$a = 1.50$	Given		$a = 1.50$	Given
$b = 2.00$	Given		$b = 2.00$	Given
$c = 2.30$	$\dfrac{\sin \alpha}{a} = \dfrac{\sin \gamma}{c}$		$c' = 0.76$	$\dfrac{\sin \alpha}{a} = \dfrac{\sin \gamma'}{c'}$
	$c = \dfrac{1.5 \sin 81°}{\sin 40°}$			$c' = \dfrac{1.5 \sin 19°}{\sin 40°}$
	$= 2.3049$			$= 0.7597$

iv. $a \ge b$

One triangle is formed (see Figure 5.10).

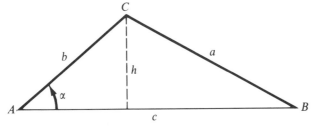

FIGURE 5.10

EXAMPLE: Let $a = 3.0$, $b = 2.0$, and $\alpha = 40°$.

Solution: $\dfrac{\sin \alpha}{a} = \dfrac{\sin \beta}{b}$

$$\dfrac{\sin 40°}{3} = \dfrac{\sin \beta}{2}$$

$$\sin \beta = \dfrac{2}{3} \sin 40°$$

$$= \dfrac{2}{3}(0.6428)$$

$$= 0.4285$$

$$\beta = 25.374$$

$\alpha = 40°$	Given
$\beta = 25°$	See above.
$\gamma = 115°$	$\gamma = 180° - \alpha - \beta$
$a = 3.0$	Given
$b = 2.0$	Given
$c = 4.2$	Since $\sin 40°/3 = \sin 115°/c$,

$$c = \dfrac{3 \sin 115°}{\sin 40°}$$

$$= 4.2299.$$

In conclusion, be sure you remember the following important rule.

When given two sides and an angle that is not an included angle, check to see whether one side is between the height of the triangle and the length of the other side. If so, there will be two solutions.

PROBLEM SET 5.3

A Problems

Solve triangle ABC in Problems 1–15. If two triangles are possible, give both solutions. If the triangle does not have a solution, tell why.

1. $a = 3.0$; $b = 4.0$; $\alpha = 125°$
2. $a = 3.0$; $b = 4.0$; $\alpha = 80°$
3. $a = 5.0$; $b = 7.0$; $\alpha = 75°$
4. $a = 5.0$; $b = 7.0$; $\alpha = 135°$
5. $a = 5.0$; $b = 4.0$; $\alpha = 125°$
6. $a = 5.0$; $b = 4.0$; $\alpha = 80°$
7. $a = 9.0$; $b = 7.0$; $\alpha = 75°$
8. $a = 9.0$; $b = 7.0$; $\alpha = 135°$
9. $a = 7.0$; $b = 9.0$; $\alpha = 52°$
10. $a = 12.0$; $b = 9.00$; $\alpha = 52.0°$
11. $a = 8.63$; $b = 11.8$; $\alpha = 47.0°$
12. $a = 10.2$; $b = 11.8$; $\alpha = 47.0°$

13. $a = 4.0; b = 5.0; \alpha = 56°$

14. $a = 4.5; b = 5.0; \alpha = 56°$

15. $a = 7.0; b = 5.0; \alpha = 56°$

B Problems

Solve triangle ABC with the parts given in Problems 16–30. If the triangle cannot be solved, tell why.

16. $a = 14.2; b = 16.3; \gamma = 35.0°$

17. $a = 14.2; b = 16.3; \gamma = 135.0°$

18. $\beta = 15.0°; \gamma = 18.0°; b = 23.5$

19. $b = 45.7; \alpha = 82.3°; \beta = 61.5°$

20. $b = 82.5; c = 52.2; \gamma = 32.1°$

21. $a = 151; b = 234; c = 416$

22. $a = 68.2; \alpha = 145°; \beta = 52.4°$

23. $\alpha = 82.5°; \beta = 16.9°; \gamma = 80.6°$

24. $a = 123; b = 225; c = 351$

25. $a = 27.2; c = 35.7; \alpha = 43.7°$

26. $c = 196; \alpha = 54.5°; \gamma = 63.0°$

27. $b = 428; c = 395; \gamma = 28.4°$

28. $a = 285; b = 209; \alpha = 42.8°$

29. $a = 612; b = 251; c = 812$

30. $b = 482.7; c = 543.9; \beta = 28.32°$

5.4 APPLICATIONS OF OBLIQUE TRIANGLES
(Which Way?)

A·E·BEARD

© DATAMATION ®

The most important skill that can be learned from the material of this chapter is the ability to select the proper trigonometric law when given a particular applied problem. In this section we will encounter applications of right triangles, the law of cosines, and the law of sines. Let's review the various types of problems we've considered.

TO SOLVE A TRIANGLE ABC

Given	Conditions on Given Information	Law to Use for Solution
1. SSS	a. The sum of the lengths of the two smaller sides is less than or equal to the length of the larger side.	No solution
	b. The sum of the lengths of the two smaller sides is greater than the length of the larger side.	Law of cosines
2. SAS	a. The angle is greater than or equal to 180°.	No solution
	b. The angle is less than 180°.	Law of cosines
3. ASA or AAS	a. The sum of the angles is greater than or equal to 180°.	No solution
	b. The sum of the angles is less than 180°.	Law of sines
4. SSA	Let α be the given angle, a and b the given sides, and h the altitude.	
	a. $\alpha \geq 90°$	
	i. $a \leq b$	No solution
	ii. $a > b$	Law of sines
	b. $\alpha < 90°$	
	i. $a < h < b$	No solution
	ii. $a = h < b$	Right-triangle solution
	iii. $h < a < b$	*Ambiguous case:* Use the law of sines to find two solutions.
	iv. $a \geq b$	Law of sines
5. AAA		No solution

Thus we use the law of cosines for cases 1 and 2, and we use the law of sines for cases 3 and 4 (remembering that, when given SSA with $h < a < b$, we should expect two solutions).

EXAMPLE: On June 22, 1970, two persons were seated exactly 200 and 285 feet, respectively, from the opposite ends of the world's longest bar, on Wharf Street in St. Louis, Missouri as shown in Figure 5.11. They calculated the angle at their location formed by the lines drawn from either end of the bar to be 85.9°. How long is the bar?

Solution: We are given SAS, so we use the law of cosines.

$$b^2 = a^2 + c^2 - 2ac \cos \beta$$
$$= 200^2 + 285^2 - 2(200)(285) \cos 85.9°$$

$= 113,074 \text{ (by calculator)}$

$b = 336.2652$

The length of the bar is 336 ft.

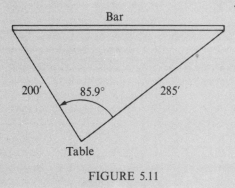

FIGURE 5.11

EXAMPLE: A boat traveling at 35.0 knots heads in the direction of S42.5°W across a large river whose current is moving at 12.0 knots in the direction of S85.3°W. Give the actual speed and direction of the boat.

Solution: This problem can be solved using vectors. We seek the direction and magnitude of the resultant vector shown in Figure 5.12.

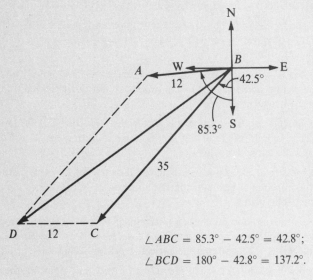

$\angle ABC = 85.3° - 42.5° = 42.8°;$

$\angle BCD = 180° - 42.8° = 137.2°.$

FIGURE 5.12

To find DB, we use the law of cosines:

$$DB^2 = 12^2 + 35^2 - 2(12)(35) \cos 137.2°$$

$$DB = 44.6 \text{ (by calculator; three significant digits).}$$

For $\angle DBC$:

$$\cos \angle DBC = \frac{44.6^2 + 35^2 - 12^2}{2(44.6)(35)}$$

$$\angle DBC = 10.5° \text{ (by calculator; three significant digits).}$$

Therefore the ship is traveling in the direction of S53.0°W (42.5° + 10.5°) with a speed of 44.6 knots.

PROBLEM SET 5.4

A Problems

1. Two sides of a parallelogram make an angle of 47°. If the lengths of the sides are 18 m and 24 m, what is the length of each diagonal?

2. Two sides of a parallelogram make an angle of 105°. If the lengths of the sides are 53 cm and 84 cm, what is the length of each diagonal?

3. An artillery-gun observer must determine the distance to a target at point T. He knows that the target is 5.20 miles from point I on a nearby island. He also knows that he (at point H) is 4.30 miles from point I. If $\angle HIT$ is 68.4°, how far is he from the target?

4. Answer the question posed in Problem 3 with $\angle HIT = 42.3°$.

5. A tree stands vertically on a hillside that makes an angle of 14° with the horizontal. If the angle of elevation of the tree at 58 feet down the hill from the base is 52°, what is the height of the tree?

6. A tree stands vertically on a hillside that makes an angle of 5° with the horizontal. If the angle of elevation of the tree at exactly 50 feet down the hill from the base is 43°, what is the height of the tree?

7. A boat moving at 18.0 knots heads in the direction of S38.2°W across a large river whose current is traveling at 5.00 knots in the direction of S13.1°W. Give the actual speed and direction of the boat.

8. A boat moving at 25.0 knots heads in the direction of N68.0°W across a river whose current is traveling at 12.0 knots in the direction of N16°W. Give the actual speed and direction of the boat.

9. A buyer is interested in purchasing a triangular lot whose vertices are LOT, but unfortunately the marker at point L has been lost. However, the deed indicates that TO is 453 ft and LO is 112 ft and that the angle at L is 82.6°. What is the distance from L to T?

10. Two forces of 220 lb and 180 lb are acting on the same point in directions that differ by 52°. What is the magnitude of the resultant, and what is the angle that it makes with each of the given forces?

B Problems

11. A UFO is sighted by people in two cities 2.300 miles apart. The UFO is between and in the same vertical plane as the cities. The angle of elevation of the UFO from the first city is 10.48° and from the second is 40.79°. At what altitude is the UFO flying, and what is the actual distance of the UFO from each city?

12. At 500 ft in the direction that the Tower of Pisa is leaning, the angle of elevation is 20.24°. If the tower leans at an angle of 5.45° from the vertical, what is the length of the tower?

BURR SHAFER

"WHAT A PITY — A BEAUTIFUL TOWER LIKE THAT COULD HAVE MADE PISA FAMOUS."

13. What is the angle of elevation of the Leaning Tower of Pisa (described in Problem 12) if you measure from a point 500 ft in the direction exactly opposite from the way it is leaning?

14. From a blimp the angle of depression to the top of the Eiffel Tower is 23.2° and to the bottom is 64.6°. After flying over the tower at the same height and at a distance of 1000 ft from the first location, you determine that the angle of depression to the top of the tower is now 31.4°. What is the height of the Eiffel Tower given that these measurements are in the same vertical plane?

15. The world's longest deep-water jetty is at Le Havre, France. Since access to the jetty is restricted, it was necessary for me to calculate its length by noting that it forms an angle of 85.0° with the shoreline. After pacing out 1000 ft along the line making an 85.0° angle with the jetty, I calculated the angle to the end of the jetty to be 83.6°. What is the length of the jetty?

16. The world's longest pier is at Hasa, Saudi Arabia. The length of the pier can be determined by measuring out a distance of 1.00 mi along a line on shore that makes an angle of 83.4° with the foot of the pier. At that point the angle to the end of the pier is 88.05°. How long is the pier?

17. Two boats leave port at the same time. The first travels 24.1 mph in the direction of S58.5°W, while the other travels 9.80 mph in the direction of N42.1°E. How far apart are the boats after two hours?

18. A sailboat is in a 5.30-mph current in the direction of S43.2°W and is being blown by a 3.20-mph wind in a direction of S25.3°W. Find the direction and speed of the sailboat.

19. A level lot has dimensions as shown in Figure 5.13. If one acre = 43,560 sq ft, what is the total cost of treating the area for poison oak if the fee is $25 per acre?

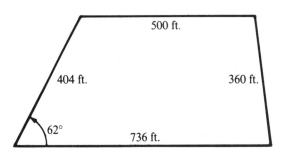

FIGURE 5.13

20. Show that the area K of a triangle ABC can be found by

$$K = \tfrac{1}{2}bc \sin \alpha = \tfrac{1}{2}ac \sin \beta = \tfrac{1}{2}ab \sin \gamma$$

when two sides and an included angle are known.

21. Show that the area K of a triangle ABC can be found by

$$K = \frac{a^2 \sin \beta \sin \gamma}{2 \sin \alpha}$$

when three angles and one side are known.

22. Use Problem 20 and the law of cosines to show that the area K of a triangle can be found by

$$K = \sqrt{s(s - a)(s - b)(s - c)}$$

when three sides are known and $s = \tfrac{1}{2}(a + b + c)$. This equation is known as *Heron's* or *Hero's Formula.*

Find the area of the triangles in Problems 23–30.

23. $a = 14.2$; $b = 16.3$; $\gamma = 35.0°$ 24. $a = 14.2$; $b = 16.3$; $\gamma = 135.0°$

25. $B = 15.0°$; $C = 18.0°$; $b = 23.5$ 26. $b = 45.7$; $\alpha = 82.3°$; $\beta = 61.5°$

27. $b = 82.5$; $c = 52.2$; $\alpha = 32.1°$ 28. $a = 151$; $b = 234$; $c = 416$

29. $a = 124$; $b = 325$; $c = 351$ 30. $a = 27.2$; $c = 35.7$; $\alpha = 43.7°$

Mind Bogglers

31. Show that the radius R of a circumscribed circle of a triangle ABC satisfies the equations

$$R = \frac{a}{2 \sin \alpha} = \frac{b}{2 \sin \beta} = \frac{c}{2 \sin \gamma}.$$

32. Show that the radius r of an inscribed circle of a triangle ABC satisfies the equation

$$r = \sqrt{\frac{(s - a)(s - b)(s - c)}{s}},$$

where $s = \frac{1}{2}(a + b + c)$.

5.5 SUMMARY AND REVIEW

TERMS

AAA (5.2)   Law of sines (5.2)
AAS (5.2)   Oblique triangle (5.1)
Ambiguous case (5.3)   SAS (5.1)
ASA (5.2)   SSA (5.3)
Law of cosines (5.1)   SSS (5.1)

CONCEPTS

(5.1) Law of cosines

$$a^2 = b^2 + c^2 - 2bc \cos \alpha \qquad \cos \alpha = \frac{b^2 + c^2 - a^2}{2bc}$$

$$b^2 = a^2 + c^2 - 2ac \cos \beta \qquad \cos \beta = \frac{a^2 + c^2 - b^2}{2ac}$$

$$c^2 = a^2 + b^2 - 2ab \cos \gamma \qquad \cos \gamma = \frac{a^2 + b^2 - c^2}{2ab}$$

(5.2) Law of sines

$$\frac{\sin \alpha}{a} = \frac{\sin \beta}{b} = \frac{\sin \gamma}{c}$$

(5.3) When given two sides and an angle that is not an included angle, check to see if one side is between the height of the triangle and the length of the other side. If so, there will be two solutions.

(5.4) Summary for solving a triangle (see page 194).

PROBLEM SET 5.5 *(Chapter 5 Test)*

(5.1–5.3) In Problems 1–5, complete the given table for triangle ABC.

	To Find	Known	Procedure	Solution
Example:			Solution:	
	β	a, b, α	$\dfrac{\sin\alpha}{a} = \dfrac{\sin\beta}{b}$	$\beta = \sin^{-1}\left(\dfrac{b\sin\alpha}{a}\right)$
1.	α	a, b, c		
2.	α	a, b, β		
3.	α	a, β, γ		
4.	b	a, α, β		
5.	b	a, c, β		

6. (5.1) Prove that $a^2 = b^2 + c^2 - 2bc\cos\alpha$.

7. (5.2) Prove that $\sin\alpha/a = \sin\gamma/c$.

8. (5.4) A mine shaft is dug into the side of a sloping hill. The shaft is dug horizontally for 485 feet. Next, a turn is made so that the angle of elevation of the second shaft is 58.0°, thus forming a 58° angle between the shafts. The shaft is then continued for 382 feet before exiting as shown in Figure 5.14. How far is it along a straight line from

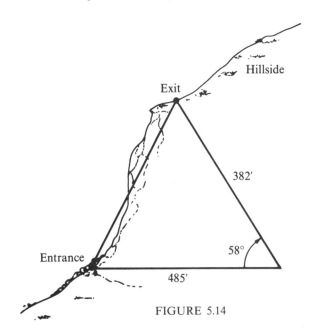

FIGURE 5.14

the entrance to the exit, assuming that all tunnels are in a single plane? If the slope of the hill follows the line from the entrance to the exit, what is the angle of elevation from the entrance to the exit?

9. (5.4) A triangularly shaped garden has two angles measuring 46.5° and 105.8°, with the side opposite the 46.5° angle measuring 38.0 m. How much fence is needed to enclose the garden?

10. (5.4) Ferndale is 7 miles N50°E of Fortuna. If I leave Fortuna at noon and travel due east at 2 mph, when will I be exactly 6 miles from Ferndale?

In mathematics there are several applications that give rise to an equation such as

$$x^2 + 1 = 0$$

that cannot be solved in the set of real numbers. A set of *complex numbers* has been defined that includes not only the real numbers but also square roots of negative numbers. Girolamo Cardano (1501–1576) worked with square roots of negative numbers as early as 1545. The number *i*, defined in Section 6.1, is called the *imaginary* unit, although it is no more "imaginary" than the number 5. When working with complex and imaginary numbers, we must keep in mind that *all* numbers are concepts and that the number *i* is just as much a number as any of the "real" numbers.

The naming of results in the history of mathematics is a strange phenomenon. In this chapter we study De Moivre's Theorem, named after Abraham de Moivre (1667–1754), who used his famous formula in Philosophical Transactions *(1707) and in* Miscellanea analytica *(1730). He was an outstanding mathematician who worked extensively with probability. However, De Moivre's Formula was really first used by Roger Cotes (1682–1716). On the other hand, De Moivre was the inventor of a result called Stirling's Formula. De Moivre was unable to receive a professorship, so he supported himself by making a London coffeehouse his headquarters and solving problems for anyone who came to him. A story about De Moivre's death is reported in Howard Eves'* In Mathematical Circles *(Boston: Prindle, Weber, & Schmidt, 1969). According to the story, De Moivre noticed that each day he required a quarter of an hour more sleep than on the preceding day. De Moivre said he would die when the arithmetic progression reached 24 hours. Finally, after sleeping longer and longer every night, he went to bed and slept for more than 24 hours and then died in his sleep.*

6

COMPLEX NUMBERS

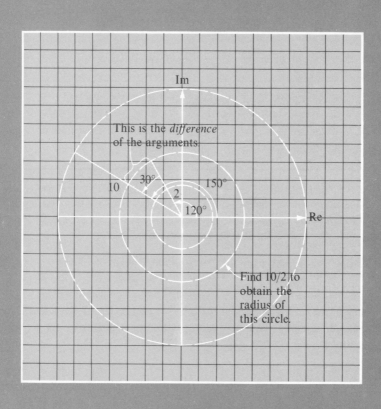

6.1 THE IMAGINARY UNIT AND COMPLEX NUMBERS *(It's All in Your Mind)*

To find the roots of certain equations, we must sometimes consider the square roots of negative numbers. Since the set of real numbers does not allow for such a possibility, we define a number that is *not a real number*. This number is $\sqrt{-1}$ and is denoted by the symbol i defined so that

$$i^2 = -1.$$

This number i is called the *imaginary unit*.

With this number we can write the square roots of any negative number as the product of a real number and the number $\sqrt{-1}$ or i. Thus we want to define i so that

$$\sqrt{-9} = \sqrt{9}\,\sqrt{-1} = 3\sqrt{-1} = 3i$$

and, for any positive real number b,

$$\sqrt{-b} = \sqrt{b}\,\sqrt{-1} = \sqrt{b}\,i.$$

We now define a new set of numbers that includes the real numbers as a subset:

Definition: The set of all numbers of the form

$$a + bi$$

(with a and b real numbers, $i = \sqrt{-1}$) is the set of *complex numbers*.

If $b = 0$, then we have the subset of the complex numbers $a + 0i = a$, which is **the set of real numbers**.

If $a = 0$ with $b \neq 0$, then we have a *pure imaginary number* $0 + bi = bi$.

If $a = 0$ with $b = 1$, then we have the *imaginary unit* $0 + 1i = i$.

We say that a complex number z is *simplified* when it is written in the form

$$z = a + bi,$$

where a and b are real numbers and $i = \sqrt{-1}$. In order to work with complex numbers, we must define equality, along with the usual arithmetic operations. Let $a + bi$ and $c + di$ be complex numbers. (Remember that a, b, c, and d are real numbers.)

Equality: $a + bi = c + di$ if and only if $a = c$ and $b = d$.

Addition: $(a + bi) + (c + di) = (a + c) + (b + d)i$.

Subtraction: $(a + bi) - (c + di) = (a - c) + (b - d)i.$

Multiplication: $(a + bi)(c + di) = (ac - bd) + (ad + bc)i.$

Division: $\dfrac{a + bi}{c + di} = \dfrac{(ac + bd) + (bc - ad)i}{c^2 + d^2} = \dfrac{ac + bd}{c^2 + d^2} + \dfrac{bc - ad}{c^2 + d^2}\,i.$

It is not necessary to memorize these definitions; we can deal with two complex numbers as we would any binomials as long as we remember that $i^2 = -1.$

EXAMPLES: Simplify each of the given complex numbers.

Solutions: "Simplify" means write in the form $a + bi.$

1. $(4 + 5i) + (3 + 4i) = 7 + 9i$

2. $(2 - i) - (3 - 5i) = -1 + 4i$

3. $(5 - 2i) + (3 + 2i) = 8$ or $8 + 0i$

4. $(4 + 3i) - (4 + 2i) = i$ or $0 + i$

5. $(2 + 3i)(4 + 2i) = 8 + 16i + 6i^2$

    $\qquad\qquad\qquad = 8 + 16i - 6$

    $\qquad\qquad\qquad = 2 + 16i$

6. $(a + bi)(c + di) = ac + adi + bci + bdi^2$
$$= ac + (ad + bc)i - bd$$
$$= (ac - bd) + (ad + bc)i$$

Notice that this result corresponds to the definition given above, but we carried out the calculations in the "usual" way while remembering that $i^2 = -1$.

7. $(4 - 3i)(4 + 3i) = 16 - 9i^2$
$$= 16 + 9$$
$$= 25$$

Example 7 gives a clue for dividing complex numbers. The definition would be difficult to remember, so instead we use the idea of *conjugates*. The complex numbers $a + bi$ and $a - bi$ are called *complex conjugates*, and each is the conjugate of the other:

$$(a + bi)(a - bi) = a^2 - b^2i^2$$
$$= a^2 + b^2,$$

which is a real number. Thus we simplify a quotient by using the conjugate of the denominator, as illustrated by the following examples.

EXAMPLES: Simplify.

1. $\dfrac{15 - 5i}{2 - i} = \dfrac{15 - 5i}{2 - i} \cdot \dfrac{2 + i}{2 + i} = \dfrac{30 + 5i - 5i^2}{4 - i^2}$

     multiply by 1.

     conjugates

$$= \dfrac{35 + 5i}{5}$$
$$= 7 + i$$

Recall that, if $10/5 = 2$, we can check by multiplying $5 \cdot 2 = 10$. For this example, we can check the division by multiplying $(2 - i)(7 + i)$:

$$(2 - i)(7 + i) = 14 - 5i - i^2$$
$$= 15 - 5i.$$

2. $\dfrac{6 + 5i}{2 + 3i} = \dfrac{6 + 5i}{2 + 3i} \cdot \dfrac{2 - 3i}{2 - 3i}$

$$= \dfrac{12 - 8i - 15i^2}{4 - 9i^2}$$

$$= \dfrac{27}{13} - \dfrac{8}{13}i$$

3. $\dfrac{a + bi}{c + di} = \dfrac{a + bi}{c + di} \cdot \dfrac{c - di}{c - di}$

$\qquad = \dfrac{ac - adi + bci - bdi^2}{c^2 - d^2 i^2}$

$\qquad = \dfrac{(ac + bd) + (bc - ad)i}{c^2 + d^2}$

Notice that this result corresponds to the definition of division given above.

PROBLEM SET 6.1

A Problems

Simplify the expressions in Problems 1–30.

1. $(3 + 3i) + (5 + 4i)$
2. $(6 - 2i) + (5 + 3i)$
3. $(4 - 2i) - (3 + 4i)$
4. $5 - (2 - 3i)$
5. $5i - (5 + 5i)$
6. $(5 - 3i) - (5 + 2i)$
7. $(3 + 4i) - (7 + 4i)$
8. $-2(-4 + 5i)$
9. $4(2 - i) - 3(-1 - i)$
10. $6(3 + 2i) + 4(-2 - 3i)$
11. $i(2 + 3i)$
12. $i(5 - 2i)$
13. $(3 - i)(2 + i)$
14. $(4 - i)(2 + i)$
15. $(5 - 2i)(5 + 2i)$
16. $(8 - 5i)(8 + 5i)$
17. $(3 - 5i)(3 + 5i)$
18. $(7 - 9i)(7 + 9i)$
19. $-i^2$
20. $-i^3$
21. i^3
22. i^4
23. $-i^4$
24. $-i^5$
25. $-i^6$
26. i^6
27. i^{11}
28. i^{236}
29. $-i^{1980}$
30. i^{1976}

B Problems

Simplify the expressions in Problems 31–44.

31. $(6 - 2i)^2$
32. $(3 + 3i)^2$
33. $(4 + 5i)^2$
34. $(3 - 5i)^3$

35. $\dfrac{-3}{1+i}$

36. $\dfrac{5}{4-i}$

37. $\dfrac{2}{i}$

38. $\dfrac{5}{i}$

39. $\dfrac{3i}{5-2i}$

40. $\dfrac{-2i}{3+i}$

41. $\dfrac{5+3i}{4-i}$

42. $\dfrac{4-2i}{3+i}$

43. $\dfrac{1-6i}{1+6i}$

44. $\dfrac{2+7i}{2-7i}$

Mind Bogglers

In Problems 45–47 let $z_1 = a + bi$, $z_2 = c + di$, and $z_3 = e + fi$.

45. Prove the commutative laws for complex numbers. That is, prove that:
$$z_1 + z_2 = z_2 + z_1;$$
$$z_1 z_2 = z_2 z_1.$$

46. Prove the associative laws for complex numbers. That is, prove that:
$$z_1 + (z_2 + z_3) = (z_1 + z_2) + z_3;$$
$$z_1(z_2 z_3) = (z_1 z_2)z_3.$$

47. Prove the distributive law for complex numbers. That is, prove that:
$$z_1(z_2 + z_3) = z_1 z_2 + z_1 z_3.$$

6.2 GRAPHICAL REPRESENTATION OF COMPLEX NUMBERS *(Paint a Picture)*

We represent the real numbers graphically by setting up a real number line such that there is a one-to-one correspondence between the real numbers and points on this line. To give a graphical representation of complex numbers, such as

$$2 + 3i, \quad -i, \quad -3 - 4i, \quad 3i, \quad -2 + \sqrt{2}i, \quad \frac{3}{2} - \frac{5}{2}i,$$

we use a two-dimensional coordinate system: the horizontal axis represents the *real axis* and the vertical axis the *imaginary axis*, so that $a + bi$ is represented by the ordered pair (a, b). Remember that a and b represent real numbers, so we plot (a, b) in the usual manner, as shown in Figure 6.1.

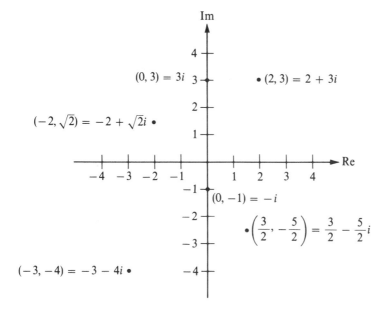

FIGURE 6.1

The coordinate system in Figure 6.1 is called the *complex plane* or the *Gaussian plane*, in honor of Karl Friedrich Gauss (1777–1855). Using this system, we can show some rather interesting and useful representations of the operations defined in the last section.

Addition of two complex numbers $a + bi$ and $c + di$ obeys the *parallelogram law*, which was developed for vectors in the last chapter. We represent the complex number $a + bi$ as a vector drawn from the origin to the point (a, b), as shown in Figure 6.2a. The sum of the complex numbers $a + bi$ and $c + di$ is the diagonal of the parallelogram formed. For subtraction we write

$$(a + bi) - (c + di) = (a + bi) + (-c - di)$$

and then use the parallelogram law as shown in Figure 6.2b.

The *absolute value* of a complex number z is, graphically, the distance between z and the origin (just as it is for real numbers). The absolute value of a complex number is also called the *modulus*. The distance formula leads to the following definition.

Definition: If $z = a + bi$, then the *absolute value* or *modulus* of z is denoted by $|z|$ and defined by

$$|z| = \sqrt{a^2 + b^2}.$$

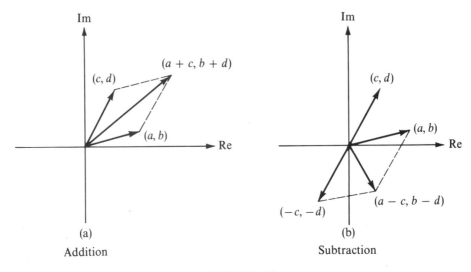

FIGURE 6.2

EXAMPLES: Find the absolute value of the following numbers.

1. $3 + 4i$. Absolute value: $|3 + 4i| = \sqrt{3^2 + 4^2}$

$$= \sqrt{25}$$

$$= 5$$

2. $-2 + \sqrt{2}i$. Absolute value: $|-2 + \sqrt{2}i| = \sqrt{4 + 2}$

$$= \sqrt{6}$$

3. -3. Absolute value: $|-3 + 0i| = \sqrt{9 + 0}$

$$= 3$$

Notice from Example 3 that the definition we've given here is consistent with the definition of absolute value given for real numbers.

The graphical representation of multiplication and division of complex numbers can be done in another, more useful way, called *trigonometric* or *polar form*. A complex number in the form $a + bi$ is said to be in *rectangular form*.

Suppose we graph a complex number as shown in Figure 6.3. Notice that the vector from the origin to (a, b) has length

$$r = \sqrt{a^2 + b^2}$$

and forms an angle θ with the real axis. We'll agree that θ, called the *argument*,

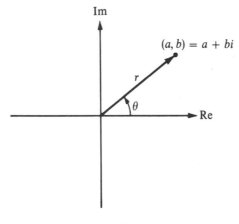

FIGURE 6.3

is chosen so that it is the smallest positive angle the terminal side makes with the real axis. From the definition of the trigonometric functions, we have the following.

$$\cos \theta = \frac{a}{r} \qquad \sin \theta = \frac{b}{r} \qquad \tan \theta = \frac{b}{a}$$

$$a = r \cos \theta \qquad b = r \sin \theta$$

Therefore

$$a + bi = r \cos \theta + ir \sin \theta$$
$$= r(\cos \theta + i \sin \theta).$$

Sometimes, $r(\cos \theta + i \sin \theta)$ is abbreviated by

$$
\begin{array}{ccccc}
r(\cos & \theta & + & i \sin & \theta) \\
\downarrow\downarrow & & & \downarrow\downarrow & \downarrow \\
r\ c & & & i\ s & \theta
\end{array}
$$

or r cis θ.

Definition: The *trigonometric* or *polar form* of a complex number $z = a + bi$ is

$$r(\cos \theta + i \sin \theta) = r \text{ cis } \theta,$$

where $r = \sqrt{a^2 + b^2}$;

$\tan \theta = b/a$ if $a \neq 0$;

$a = r \cos \theta$;

$b = r \sin \theta$.

This representation is unique for $0 \leq \theta < 360°$ for all z except $0 + 0i$.

EXAMPLES: Change the complex numbers to trigonometric form.

1. $1 - \sqrt{3}i$

    Solution: $a = 1$ and $b = -\sqrt{3}$; the number is in quad IV.

    $$r = \sqrt{1^2 + (-\sqrt{3})^2} \qquad \tan \theta = \frac{-\sqrt{3}}{1}$$

    $$= \sqrt{4} \qquad\qquad\qquad = -\sqrt{3} \quad \text{The reference angle is } 60°;$$

    $$= 2 \qquad\qquad\qquad\qquad\qquad\qquad \text{in quad IV: } \theta = 300°.$$

    Thus, $1 - \sqrt{3}i, = 2 \text{ cis } 300°$.

2. $-3 - 5i$

    Solution: $a = -3$ and $b = -5$; the number is in quad III.

    $$r = \sqrt{(-3)^2 + (-5)^2} \qquad \tan \theta = \frac{-5}{-3} \quad \text{The reference angle is } 59°;$$

    $$= \sqrt{9 + 25} \qquad\qquad\qquad\qquad\qquad \text{in quad III: } \theta = 239°.$$

    $$= \sqrt{34}$$

    Thus, $-3 - 5i = \sqrt{34} \text{ cis } 239°$.

3. $4.310 + 5.516i$

    Solution: $a = 4.310$ and $b = 5.516$; the number is in quad I.

    $$r = \sqrt{(4.310)^2 + (5.516)^2} \qquad \tan \theta = \frac{5.516}{4.310}$$

    $$= \sqrt{49} \qquad\qquad\qquad\qquad = 1.2798$$

    $$= 7 \qquad\qquad\qquad\qquad\qquad \theta = 52°$$

    Thus, $4.310 + 5.516i = 7 \text{ cis } 52°$.

4. $6i$

    Solution: $a = 0$ and $b = 6$. By inspection, $6i = 6 \text{ cis } 90°$. (Notice that $\tan \theta$ is not defined for $\theta = 90°$.)

EXAMPLES: Change the complex number to rectangular form.

1. $4 \text{ cis } 330°$

    Solution: $r = 4$ and $\theta = 330°$.

    $$a = 4 \cos 330° \qquad b = 4 \sin 330°$$

    $$= 4\left(\frac{\sqrt{3}}{2}\right) \qquad = 4\left(-\frac{1}{2}\right)$$

    $$= 2\sqrt{3} \qquad\qquad = -2$$

    Thus, $4 \text{ cis } 330° = 2\sqrt{3} - 2i$.

2. $5(\cos 38° + i \sin 38°)$

Solution: $r = 5$ and $\theta = 38°$.

$$a = 5 \cos 38° \qquad b = 5 \sin 38°$$
$$= 3.94 \qquad\qquad = 3.08$$

Thus, $5(\cos 38° + i \sin 38°) = 3.94 + 3.08i$.

In the next section we'll continue our discussion of graphical representations of multiplication and division.

PROBLEM SET 6.2

A Problems

In Problems 1–6 plot each pair of given numbers, and show the sum graphically as well as algebraically.

1. $3 + i, 5 + 6i$ 2. $7 - i, 2 - 5i$

3. $3 + 2i, -2 + 5i$ 4. $-3 - 2i, 3 - 5i$

5. $-1 + 3i, 4 - i$ 6. $2 + 4i, -1 + i$

In Problems 7–12 plot each pair of numbers, and show the difference graphically as well as algebraically.

7. $3 + i, 5 + 6i$ 8. $7 - i, 2 - 5i$

9. $3 + 2i, -2 + 5i$ 10. $-3 - 2i, 3 - 5i$

11. $-1 + 3i, 4 - i$ 12. $2 + 4i, -1 + i$

B Problems

Plot and then change each of the numbers in Problems 13–28 to polar form. Use exact values whenever possible.

13. $1 + i$ 14. $1 - i$

15. $\sqrt{3} - i$ 16. $\sqrt{3} + i$

17. $1 - \sqrt{3}i$ 18. $-1 - \sqrt{3}i$

19. $\sqrt{12} - 2i$ 20. $-\sqrt{6} + \sqrt{2}i$

21. 1 22. 5

23. $-i$ 24. $-4i$

25. $1.7207 + 2.4575i$ 26. $5.7956 - 1.5529i$

27. $-0.6946 + 3.9392i$ 28. $1.5321 - 1.2856i$

Plot and then change each of the numbers in Problems 29–44 to rectangular form. Use exact values whenever possible.

29. $2(\cos 45° + i \sin 45°)$ 30. $3(\cos 60° + i \sin 60°)$

31. $4(\cos 315° + i \sin 315°)$ 32. $5(\cos 4\pi/3 + i \sin 4\pi/3)$

33. $\cos 5\pi/6 + i \sin 5\pi/6$ 34. $5 \operatorname{cis} 3\pi/2$

35. $\operatorname{cis} 0$ 36. $4 \operatorname{cis} 30°$

37. $3 \operatorname{cis} 2\pi/3$ 38. $2 \operatorname{cis} \pi$

39. $10 \operatorname{cis} 65°$ 40. $8 \operatorname{cis} 24°$

41. $7 \operatorname{cis} 135°$ 42. $6 \operatorname{cis} 247°$

43. $9 \operatorname{cis} 190°$ 44. $10 \operatorname{cis} 371°$

Mind Bogglers

45. If $z = a + bi$ and $\bar{z} = a - bi$, show that

$$|z| = \sqrt{z \cdot \bar{z}}.$$

46. If $z_1 = a + bi$ and $z_2 = c + di$, show that

$$|z_1 + z_2| \le |z_1| + |z_2|.$$

This relationship is called the *triangle inequality*.

6.3 OPERATIONS IN POLAR FORM *(A Complex Look at Trig)*

In the last section we began to look at the geometric interpretation of the operations with complex numbers. For multiplication and division we found a very useful trigonometric form of complex numbers.

Consider the number

$$2(\cos 15° + i \sin 15°) \cdot 3(\cos 30° + i \sin 30°)$$
$$= (1.9319 + 0.5176i)(2.5981 + 1.5i).$$

Algebraic (rectangular form):

$$(1.9319 + 0.5176i)(2.5981 + 1.5i) = 5.0191 + 2.8979i + 1.3448i + 0.7764i^2$$
$$= 4.2427 + 4.2427i$$

Trigonometric (polar form):

$$2(\cos 15° + i \sin 15°) \cdot 3(\cos 30° + i \sin 30°) = 2 \cdot 3(\cos 15° + i \sin 15°)(\cos 30° + i \sin 30°)$$
$$= 6(\cos 15° \cos 30° + \cos 15° \sin 30°\, i$$
$$+ \sin 15° \cos 30°\, i + \sin 15° \sin 30°\, i^2)$$
$$= 6[(\cos 15° \cos 30° - \sin 15° \sin 30°)$$
$$+ (\cos 15° \sin 30° + \sin 15° \sin 30°)i]$$

$$= 6[\cos(15° + 30°) + \sin(15° + 30°)]$$
$$= 6(\cos 45° + \sin 45°)$$

Notice that the argument of the answer in trigonometric form is simply the sum of the arguments of the two given numbers! This result holds in general (you are asked to show this in the problems) and leads to a geometric interpretation of multiplication.

Let $z_1 = r_1 \operatorname{cis} \theta_1$ and $z_2 = r_2 \operatorname{cis} \theta_2$ be complex numbers. Then,

$$z_1 z_2 = r_1 r_2 \operatorname{cis}(\theta_1 + \theta_2).$$

Graphical: $2 \operatorname{cis} 15° \cdot 3 \operatorname{cis} 30°$. The product is found by drawing an angle representing the sum of the arguments on a circle whose radius is the product of the moduli as shown in Figure 6.4.

This theorem for the multiplication of complex numbers in trigonometric form is extremely useful, as the following examples illustrate.

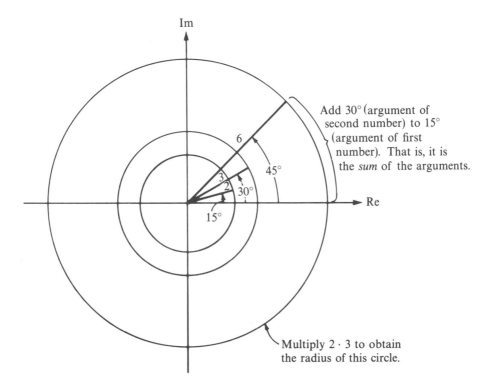

FIGURE 6.4

EXAMPLES: Simplify.

1. $5 \text{ cis } 38° \cdot 4 \text{ cis } 75° = 5 \cdot 4 \text{ cis}(38° + 75°)$
$$= 20 \text{ cis } 113°$$

2. $\sqrt{2} \text{ cis } 188° \cdot 2\sqrt{2} \text{ cis } 310° = 4 \text{ cis } 498°$
$$= 4 \text{ cis } 138°$$

3. $(2 \text{ cis } 48°)^3 = (2 \text{ cis } 48°)(2 \text{ cis } 48°)^2$
$$= (2 \text{ cis } 48°)(4 \text{ cis } 96°)$$
$$= 8 \text{ cis } 144°$$

   Notice that this result is the same as
$$(2 \text{ cis } 48°)^3 = 2^3 \text{ cis } 3 \cdot 48°$$
$$= 8 \text{ cis } 144°.$$

4. $(1 - \sqrt{3}i)^5$. We first change to trigonometric form.

   $a = 1; b = -\sqrt{3}$; quad IV.

   $r = \sqrt{1 + 3}$ $\tan \theta = -\sqrt{3}/1$

   $\quad = 2$ $\theta = 300°$ (quad IV)

   $(1 - \sqrt{3})^5 = (2 \text{ cis } 300°)^5$
$$= 2^5 \text{ cis } 5 \cdot 300°$$
$$= 32 \text{ cis } 1500°$$
$$= 32 \text{ cis } 60°$$

   If we want the answer in rectangular form, we can now change back.

   $a = 32 \cos 60°$ $b = 32 \sin 60°$

   $\quad = 32(\frac{1}{2})$ $\quad = 32(\frac{1}{2}\sqrt{3})$

   $\quad = 16$ $\quad = 16\sqrt{3}$

   Thus, $(1 - \sqrt{3})^5 = 16 + 16\sqrt{3}i$.

As you can see from the last example, multiplication in polar form extends quite nicely for any positive integral power.

If n is a natural number, then
$$(r \text{ cis } \theta)^n = r^n \text{ cis } n\theta$$
for a complex number $r \text{ cis } \theta = r(\cos \theta + i \sin \theta)$.

We will generalize this result in the next section.

For division, consider the number

$$10(\cos 150° + i \sin 150°) \div 2(\cos 120° + i \sin 120°)$$
$$= (-5\sqrt{3} + 5i) \div (-1 + \sqrt{3}i)$$

Algebraic (rectangular form):

$$\frac{-5\sqrt{3} + 5i}{-1 + \sqrt{3}i} = \frac{-5\sqrt{3} + 5i}{-1 + \sqrt{3}i} \cdot \frac{-1 - \sqrt{3}i}{-1 - \sqrt{3}i}$$

$$= \frac{5\sqrt{3} + 15i - 5i - 5\sqrt{3}i^2}{1 - 3i^2}$$

$$= \frac{10\sqrt{3} + 10i}{4}$$

$$= \frac{5}{2}\sqrt{3} + \frac{5}{2}i$$

Trigonometric (polar form):

$$\frac{10(\cos 150° + i \sin 150°)}{2(\cos 120° + i \sin 120°)}$$

$$= 5 \frac{\cos 150° + i \sin 150°}{\cos 120° + i \sin 120°} \cdot \frac{\cos 120° - i \sin 120°}{\cos 120° - i \sin 120°}$$

$$= 5 \frac{\cos 150° \cos 120° - i \cos 150° \sin 120° + i \sin 150° \cos 120° - i^2 \sin 150° \sin 120°}{\cos^2 120° - i^2 \sin^2 120°}$$

$$= 5 \frac{(\cos 150° \cos 120° + \sin 150° \sin 120°) + i(\sin 150° \cos 120° - \cos 150° \sin 120°)}{1}$$

$$= 5[\cos(150° - 120°) + i \sin(150° - 120°)]$$

$$= 5(\cos 30° + i \sin 30°)$$

Graphical: The quotient of two complex numbers is found by drawing an angle representing the difference of the respective arguments on a circle whose radius is the quotient of the respective moduli, as shown in Figure 6.5.

The above results lead to a formula for the quotient of any complex numbers in trigonometric form. The proof is left as a problem.

Let $z_1 = r_1 \operatorname{cis} \theta_1$ and $z_2 = r_2 \operatorname{cis} \theta_2$ be nonzero complex numbers. Then,

$$\frac{z_1}{z_2} = \frac{r_1}{r_2} \operatorname{cis}(\theta_1 - \theta_2).$$

EXAMPLE: Find

$$\frac{15(\cos 48° + i \sin 48°)}{5(\cos 125° + i \sin 125°)}.$$

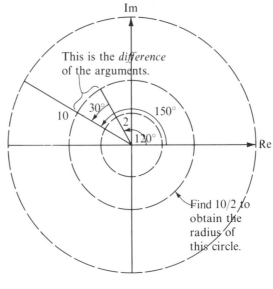

FIGURE 6.5

Solution:

$$\frac{15 \text{ cis } 48^\circ}{5 \text{ cis } 125^\circ} = 3 \text{ cis}(48^\circ - 125^\circ)$$
$$= 3 \text{ cis}(-77^\circ)$$
$$= 3 \text{ cis } 283^\circ$$

PROBLEM SET 6.3

A Problems

Perform the indicated operations in Problems 1–19.

1. $2 \text{ cis } 60^\circ \cdot 3 \text{ cis } 150^\circ$

2. $3 \text{ cis } 48^\circ \cdot 5 \text{ cis } 92^\circ$

3. $4(\cos 65^\circ + i \sin 65^\circ) \cdot 12(\cos 87^\circ + i \sin 87^\circ)$

4. $7(\cos 83^\circ + i \sin 83^\circ) \cdot 8(\cos 12^\circ + i \sin 12^\circ)$

5. $\dfrac{8 \text{ cis } 30^\circ}{4 \text{ cis } 15^\circ}$

6. $\dfrac{12 \text{ cis } 250^\circ}{4 \text{ cis } 120^\circ}$

7. $\dfrac{5(\cos 315^\circ + i \sin 315^\circ)}{2(\cos 48^\circ + i \sin 48^\circ)}$

8. $\dfrac{9(\cos 87^\circ + i \sin 87^\circ)}{5(\cos 28^\circ + i \sin 28^\circ)}$

9. $\dfrac{20(\cos 40^\circ + i \sin 40^\circ)}{10(\cos 210^\circ + i \sin 210^\circ)}$

10. $\dfrac{18(\cos 25^\circ + i \sin 25^\circ)}{9(\cos 135^\circ + i \sin 135^\circ)}$

11. $6(\cos 215° + i \sin 215°) \cdot 3(\cos 312° + i \sin 312°)$

12. $\sqrt{5}(\cos 125° + i \sin 125°) \cdot \sqrt{45}(\cos 312° + i \sin 312°)$

13. $(2 \text{ cis } 50°)^3$

14. $(3 \text{ cis } 60°)^4$

15. $(\cos 210° + i \sin 210°)^5$

16. $(\cos 85° + i \sin 85°)^9$

17. $(2 - 2i)^4$

18. $(\sqrt{3} - i)^8$

19. $(1 + i)^6$

B Problems

Perform the indicated operations in Problems 20–29. Leave your answers in polar form.

20. $(1 + i)(\sqrt{3} - i)$

21. $\dfrac{1 + i}{\sqrt{3} - i}$

22. $\dfrac{1 - \sqrt{3}i}{\sqrt{3} + i}$

23. $(1 - \sqrt{3}i)(\sqrt{3} + i)$

24. $(2 + 2i)(1 + i)^2$

25. $3i(4 - 4i)$

26. $4i(5 - 5i)(4 - 4i)(3 - 3i)$

27. $(1 + i)^5$

28. $(1 + i\sqrt{3})^4$

29. $\dfrac{(1 + i)^2}{1 + \sqrt{3}i}$

Work each of Problems 30–45 using (a) rectangular form, (b) polar form, and (c) graphical form.

30. $2(\cos 60° + i \sin 60°) \cdot 3(\cos 210° + i \sin 210°)$

31. $3(\cos 300° + i \sin 300°) \cdot 4(\cos 30° + i \sin 30°)$

32. $8(\cos 45° + i \sin 45°) \cdot 2(\cos 150° + i \sin 150°)$

33. $(3\sqrt{3} - 3i)(1 + i)$

34. $(2 + 2i)(1 - \sqrt{3}i)$

35. $3i(\sqrt{3} + i)$

36. $\dfrac{4(\cos 60° + i \sin 60°)}{2(\cos 210° + i \sin 210°)}$

37. $\dfrac{12(\cos 300° + i \sin 300°)}{4(\cos 30° + i \sin 30°)}$

38. $\dfrac{8(\cos 45° + i \sin 45°)}{2(\cos 150° + i \sin 150°)}$

39. $\dfrac{3\sqrt{3} - 3i}{1 + i}$

40. $\dfrac{2 + 2i}{1 - \sqrt{3}i}$

41. $\dfrac{3i}{\sqrt{3} + i}$

42. $(5 \text{ cis } 48°)(3 \text{ cis } 25°)$

43. $(-3.2253 + 8.4022i)(3.4985 + 1.9392i)$

44. $\dfrac{7 \text{ cis } 135°}{5 \text{ cis } 53°}$

45. $\dfrac{-3.2253 + 8.4022i}{3.4985 + 1.9392i}$

46. Prove that

$$r_1 \text{ cis } \theta_1 \cdot r_2 \text{ cis } \theta_2 = r_1 r_2 \text{ cis}(\theta_1 + \theta_2).$$

47. Prove that

$$\frac{r_1 \text{ cis } \theta_1}{r_2 \text{ cis } \theta_2} = \frac{r_1}{r_2} \text{ cis}(\theta_1 - \theta_2).$$

Mind Bogglers

48. If

$$\cos \theta = 1 - \frac{\theta^2}{2!} + \frac{\theta^4}{4!} - \frac{\theta^6}{6!} + \cdots + \frac{(-1)^n \theta^{2n}}{(2n)!} + \cdots,$$

$$\sin \theta = \theta - \frac{\theta^3}{3!} + \frac{\theta^5}{5!} - \frac{\theta^7}{7!} + \cdots + \frac{(-1)^n \theta^{2n+1}}{(2n+1)!} + \cdots,$$

and

$$e^{i\theta} = 1 + (i\theta) + \frac{(i\theta)^2}{2!} + \frac{(i\theta)^3}{3!} + \frac{(i\theta)^4}{4!} + \cdots + \frac{(i\theta)^n}{n!} + \cdots,$$

show that $e^{i\theta} = \cos \theta + i \sin \theta$. This equation is called *Euler's Formula*.

49. Using Problem 48, show that $e^{i\pi} = -1$.

6.4 DE MOIVRE'S FORMULA *(Getting at the Root of the Problem)*

In the last section we used

$$(r \text{ cis } \theta)^n = r^n \text{ cis } n\theta$$

for n a natural number. However, we also noted that this result is true for any multiple of 360°, so that we could write expressions such as

$$\text{cis } 1500° = \text{cis } 60°.$$

That is, we could write

$$(r \text{ cis } \theta)^n = r^n \text{ cis } n(\theta + 360° \, k)$$

where k is any integer. As long as we restrict n to a natural number, this second form merely gives the same values as the formula of the last section. However, our intent is to generalize the result by allowing n to be any real number. Making this generalization gives a result which is called *De Moivre's Formula*, and is stated without proof.

De Moivre's Formula. If n is any real number,

$$(r \text{ cis } \theta)^n = r^n \text{ cis } n(\theta + 360°k)$$

for any integer k.

EXAMPLES:

1. Find the square roots of
$$-\frac{9}{2} + \frac{9}{2}\sqrt{3}i.$$

Solution: First change to polar form.

$$r = \sqrt{\left(-\frac{9}{2}\right)^2 + \left(\frac{9}{2}\sqrt{3}\right)^2}$$

$$= \sqrt{\frac{81}{4} + \frac{81 \cdot 3}{4}}$$

$$= \sqrt{81\left(\frac{1}{4} + \frac{3}{4}\right)}$$

$$= 9$$

$$\tan \theta = \frac{\frac{9}{2}\sqrt{3}}{-\frac{9}{2}}$$

$$= -\sqrt{3}$$

$$\theta = 120° \text{ (quad II)}$$

$$\sqrt{-\frac{9}{2} + \frac{9}{2}\sqrt{3}i} = \left(-\frac{9}{2} + \frac{9}{2}\sqrt{3}i\right)^{1/2}$$

$$= (9 \text{ cis } 120°)^{1/2}$$

$$= 9^{1/2} \text{ cis}\left(\frac{120° + 360°k}{2}\right) \qquad \text{By De Moivre's Formula}$$

$$= 3 \text{ cis}(60° + 180°k)$$

$$k = 0: \ 3 \text{ cis } 60° \ = \frac{3}{2} + \frac{3}{2}\sqrt{3}i$$

$$k = 1: \ 3 \text{ cis } 240° = -\frac{3}{2} - \frac{3}{2}\sqrt{3}i$$

All other integral values of k repeat one of the previously found roots. For example,

$$k = 2: \ 3 \text{ cis } 420° = \frac{3}{2} + \frac{3}{2}\sqrt{3}i.$$

Check:

$$\left(\frac{3}{2} + \frac{3}{2}\sqrt{3}i\right)^2 = \frac{9}{4} + \frac{9}{2}\sqrt{3}i + \frac{9}{4} \cdot 3i^2$$

$$= -\frac{9}{2} + \frac{9}{2}\sqrt{3}i$$

$$\left(-\frac{3}{2} - \frac{3}{2}\sqrt{3}i\right)^2 = \frac{9}{4} + \frac{9}{2}\sqrt{3}i + \frac{9}{4} \cdot 3i^2$$

$$= -\frac{9}{2} + \frac{9}{2}\sqrt{3}i$$

Notice that we've found two square roots, as expected. In fact, a complex number will have n nth roots.

2. Find the five fifth roots of 32.

    Solution: $32 = 32 \text{ cis } 0°$. Thus

    $$\sqrt[5]{32} = (32)^{1/5} = 32^{1/5} \text{ cis}\left(\frac{0° + 360°k}{5}\right)$$

↑	↑

    This represents   This is r, which represents a length
    all roots of 32.   and therefore is the positive real
                           root.
                               ↓

    $$= 2 \text{ cis } 72°k.$$

    $k = 0:\ 2 \text{ cis } 0°\ \ = 2$
    $k = 1:\ 2 \text{ cis } 72° = 0.6180 + 1.9021i$
    $k = 2:\ 2 \text{ cis } 144° = -1.6180 + 1.1756i$
    $k = 3:\ 2 \text{ cis } 216° = -1.1680 - 1.1756i$
    $k = 4:\ 2 \text{ cis } 288° = 0.6180 - 1.9021i$

All other integral values for k repeat those listed here.

If we represent the fifth roots of 32 graphically, as shown in Figure 6.6, we notice that they all lie on a circle of radius 2 and are equally spaced.

If n is a positive integer, then

$$(a + bi)^{1/n} = (r \text{ cis } \theta)^{1/n},$$

and the roots are equally spaced on the circle of radius r centered at the origin.

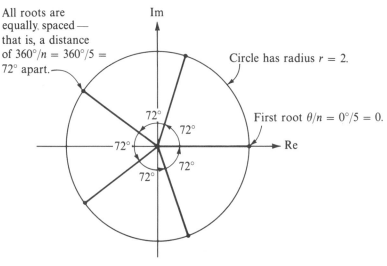

All roots are equally spaced — that is, a distance of $360°/n = 360°/5 = 72°$ apart.

Circle has radius $r = 2$.

First root $\theta/n = 0°/5 = 0$.

FIGURE 6.6

The first root, which is located so that its argument is θ/n, is called the principal nth root.

PROBLEM SET 6.4

A Problems

Find the indicated roots of the numbers in Problems 1–20. Leave your answers in trigonometric form.

1. square roots of 9 cis 240°
2. square roots of 16 cis 100°
3. square roots of 25 cis 300°
4. cube roots of 8 cis 240°
5. cube roots of 27 cis 300°
6. cube roots of 64 cis 216°
7. fourth roots of 16 cis 352°
8. fourth roots of 81 cis 88°
9. fourth roots of 256 cis 224°
10. fifth roots of 32 cis 200°
11. fifth roots of 32 cis 160°
12. fifth roots of 243 cis 60°
13. cube roots of -1
14. cube roots of 8
15. cube roots of 27
16. fourth roots of i
17. fourth roots of $-1 - i$
18. fourth roots of $1 + i$
19. fifth roots of $-i$
20. fifth roots of 1

B Problems

Find the indicated roots of the numbers in Problems 21–26. Leave your answers in rectangular form. Also, show the roots graphically.

21. cube roots of 1
22. square roots of $\dfrac{25}{2} - \dfrac{25\sqrt{3}}{2}i$
23. cube roots of -8
24. fourth roots of 1
25. cube roots of $4\sqrt{3} - 4i$
26. fourth roots of $12.2567 + 10.2846i$

Find the indicated roots of the numbers in Problems 27–34.

27. sixth roots of -64
28. sixth roots of $64i$
29. ninth roots of 1
30. ninth roots of $-1 + i$
31. tenth roots of i
32. tenth roots of 1
33. $(4\sqrt{2} + 4\sqrt{2}i)^{2/3}$
34. $(-16 + 16\sqrt{3}i)^{3/5}$

Mind Bogglers

35. Solve $x^5 - 1 = 0$.
36. Solve $x^4 + x^3 + x^2 + x + 1 = 0$. (Hint: see Problem 35.)

6.5 POLAR COORDINATES *(A Trip to the Pole)*

In the first part of this book we used a rectangular coordinate system. Now we'll consider a different coordinate system, called the *polar coordinate system.* In this system we fix a point O, called the *origin* or *pole*, and measure a point in the plane by an ordered pair $P(r, \theta)$, where θ measures the angle from the positive x-axis and r represents the directed distance from the pole to the point P.

When plotting points, we generally first measure an angle θ and then measure out a length r along the directed line segment **OP**. If θ is positive, the angle is measured in a counterclockwise direction. If r is positive, it is measured along the directed line segment **OP**; if it is negative, the distance is measured along the directed line segment $-$**OP**. The polar coordinate system is shown in Figure 6.7.

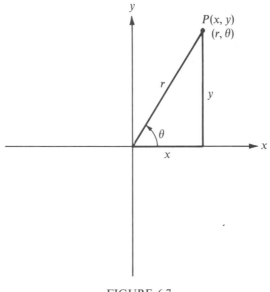

FIGURE 6.7

EXAMPLE: Plot the following points in polar coordinates:

$$(2, \pi/4), (2, -\pi/4), (-2, \pi/4), (-2, -\pi/4), (3, 0), (4, \pi), \text{ and } (1, 1).$$

Solution: The points are plotted on special graph paper called *polar graph paper*, as shown in Figure 6.8. If you don't have polar graph paper, you can get by with a compass and protractor. Notice that all the angles are measured in radians.

The relationship between the two coordinate systems can easily be found by using the definition of the trigonometric functions (see Figure 6.7).

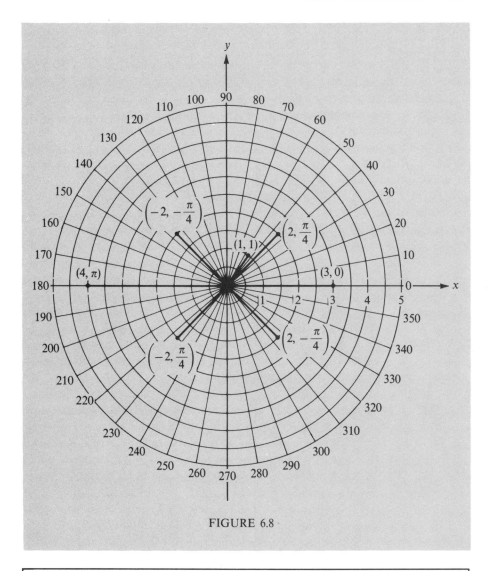

FIGURE 6.8

1. To change from polar to rectangular:

    $x = r \cos \theta;$

    $y = r \sin \theta.$

2. To change from rectangular to polar:

    $r = \sqrt{x^2 + y^2}$

    $\tan \theta = y/x.$

Notice that these relationships are the same as those used in Section 6.2.

If r and θ are connected by an equation, we can speak of the graph of the equation. For example, $r = 5$ is the equation of a circle with center at the origin and radius 5. Also, the equation $\theta = \pi/3$ is the equation of a line passing through the origin (see Figure 6.9).

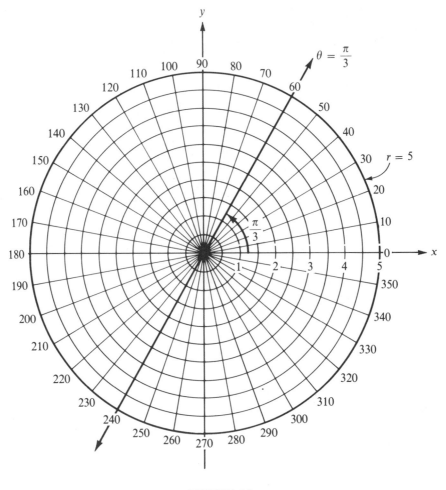

FIGURE 6.9

In rectangular coordinates every ordered pair (x, y) is associated in a one-to-one fashion with the points in the plane. In polar coordinates every ordered pair (r, θ) is *not* associated in a one-to-one fashion with the points in the plane.

EXAMPLES:

1. Given an ordered pair

    Rectangular coordinates, (4, 3)   Polar coordinates, $\left(2, \dfrac{\pi}{6}\right)$

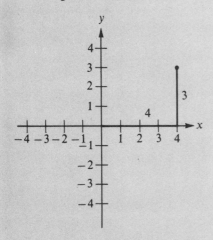

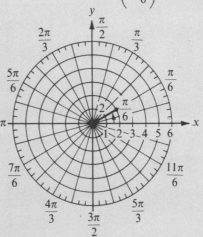

There is exactly one point in the plane.   *There is exactly one point in the plane.*

2. Given a point in the plane

    Rectangular coordinates   Polar coordinates

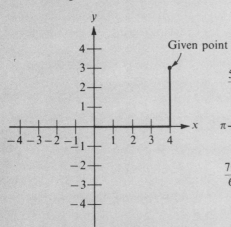

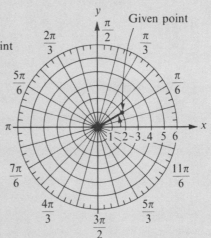

There is exactly one ordered pair   *There are many ordered pairs*
describing this point—namely, (4, 3).   *describing this point:* $(2, \pi/6),$
                                              $(2, 2\pi + \pi/6), (2, 4\pi + \pi/6), \ldots,$
                                              $(-2, \pi + \pi/6), (-2, 3\pi + \pi/6),$
                                              $(-2, 5\pi + \pi/6), \ldots$

Notice that, if we are given a point (r, θ) (other than the pole) with polar co-ordinates, then $(-r, \pi + \theta)$ also represents that point; (r, θ) and $(-r, \pi + \theta)$ are the two *primary representations of the point*. There are infinitely many others, all of which are multiples of 2π added to the primary representations. Thus,

$$(r, \theta) \qquad\qquad (-r, \theta + \pi)$$
$$(r, \theta + 2\pi) \qquad\qquad (-r, \theta + 3\pi)$$
$$(r, \theta - 2\pi) \qquad\qquad (-r, \theta - \pi)$$
$$\vdots \qquad\qquad\qquad \vdots$$
$$(r, \theta + 2k\pi) \qquad\qquad (-r, \theta + (2k + 1)\pi)$$

all name the same point for any integer k.

EXAMPLES: Show that the given points lie on the graph of

$$r = \frac{2}{1 - \cos \theta}.$$

1. $(2, \pi/2)$

    Solution: Substituting,

    $$2 \overset{?}{=} \frac{2}{1 - \cos(\pi/2)}$$
    $$\overset{?}{=} \frac{2}{1 - 0};$$
    $$2 = 2.$$

    Thus the point is on the curve, since it satisfies the equation.

2. $(-2, 3\pi/2)$

    Solution: Substituting,

    $$-2 \overset{?}{=} \frac{2}{1 - \cos(3\pi/2)}$$
    $$\overset{?}{=} \frac{2}{1 - 0};$$
    $$-2 \neq 2.$$

    The equation is not satisfied, but we *cannot* say that the point is not on the curve. Indeed, we see from Example 1 that it is on the curve, since $(-2, 3\pi/2)$ and $(2, \pi/2)$ name the same point! So even if one primary representation of a point does not satisfy the equation, we must still check the other one.

3. $(-1, 2\pi)$

    Solution: Substituting,

    $$-1 \overset{?}{=} \frac{2}{1 - \cos 2\pi}$$

$$\stackrel{?}{=} \frac{2}{1-1}, \text{ which is undefined.}$$

Checking the other representation of the same point—namely, $(1, 3\pi)$:

$$1 \stackrel{?}{=} \frac{2}{1 - \cos 3\pi}$$

$$\stackrel{?}{=} \frac{2}{1 - (-1)};$$

$$1 = 1.$$

Thus the point is on the curve.

The above examples lead us to the following definition.

Definition: A point other than the pole is on a polar-form curve if and only if at least one of its primary representations satisfies the given equation.

PROBLEM SET 6.5

A Problems

In Problems 1–10 plot the points and give the rectangular coordinates.

1. $(4, \pi/4)$ 2. $(6, \pi/3)$ 3. $(5, 2\pi/3)$ 4. $(3, -\pi/6)$ 5. $(3/2, -5\pi/6)$

6. $(5, -\pi/2)$ 7. $(-4, \pi/6)$ 8. $(-1, -2\pi/3)$ 9. $(-4, \pi/4)$ 10. $(-3, \pi/6)$

In Problems 11–20 plot the points and give the polar coordinates. Use exact values wherever possible.

11. $(5, 5)$ 12. $(-1, \sqrt{3})$ 13. $(2, -2\sqrt{3})$ 14. $(-2, -2)$ 15. $(3, -3)$

16. $(-6, 6)$ 17. $(-\sqrt{3}, 1)$ 18. $(4, 3)$ 19. $(-12, 5)$ 20. $(3, 7)$

B Problems

21. Derive the equations for changing from polar coordinates to rectangular coordinates.

22. Derive the equations for changing from rectangular coordinates to polar coordinates.

In Problems 23–27 tell whether the given points lie on the curve $r = 5/(1 - \sin \theta)$.

23. $(10, \pi/6)$ 24. $(5, \pi/2)$ 25. $(-10, 5\pi/6)$ 26. $(-10/3, 5\pi/6)$

27. $(20 + 10\sqrt{3}, \pi/3)$

In Problems 28–32 tell whether the given points lie on the curve $r = 2(1 - \cos \theta)$.

28. $(1, \pi/3)$ 29. $(1, -\pi/3)$ 30. $(-1, \pi/3)$ 31. $(-2, \pi/2)$

32. $(-2 - \sqrt{2}, \pi/4)$

Find three ordered pairs satisfying each of the equations in Problems 33–44. Give both primary representations for each point.

33. $r^2 = 9 \cos \theta$ 34. $r^2 = 9 \cos 2\theta$ 35. $r = 3\theta$

36. $r = 5\theta$ 37. $r\theta = 4$ 38. $r = 2 - 3 \sin \theta$

39. $r = 2(1 + \cos \theta)$ 40. $r = 2 \cos \theta$ 41. $r = \tan \theta$

42. $r = 6 \cos 3\theta$ 43. $r = \dfrac{8}{1 - 2 \cos \theta}$ 44. $\dfrac{r}{1 - \sin \theta} = 2$

45. What is the distance between $(3, \pi/3)$ and $(7, \pi/4)$? (Hint: use the law of cosines.)

6.6 GRAPHING POLAR CURVES *(Negotiate the Curve)*

Basically, we sketch polar-form curves by plotting points. We begin by selecting values for θ from 0 to 2π that are exact values and then find the corresponding r values. However, if the curve is symmetric with respect to the

x-axis, we check values of θ from 0 to π;
y-axis, we check values of θ from $-\pi/2$ to $\pi/2$;
origin, we check values of θ from 0 to π.

If any two of the three symmetries hold, then the third will necessarily hold. In this case we check values of θ from 0 to $\pi/2$.

The rules for determining symmetry require that, whenever one representation of a point does not satisfy the test, we must check the other representation.

A Polar-Form Equation Is Symmetric with Respect to:

1. the *x*-axis if the equation remains unchanged when we substitute

$(r, -\theta)$ or $(-r, \pi - \theta)$ for (r, θ);

2. the *y*-axis if the equation remains unchanged when we substitute

$(r, \pi - \theta)$ or $(-r, 2\pi - \theta)$ for (r, θ);

3. the origin if the equation remains unchanged when we substitute

$(r, \pi + \theta)$ or $(-r, 2\pi + \theta)$ for (r, θ).

The following results from trigonometry might be helpful when you are testing for symmetry and graphing polar-form curves:

$$\cos(-\theta) = \cos\theta \qquad \cos(\pi - \theta) = -\cos\theta \qquad \cos(\pi + \theta) = -\cos\theta \qquad \cos(\theta \pm 2\pi) = \cos\theta$$
$$\sin(-\theta) = -\sin\theta \qquad \sin(\pi - \theta) = \sin\theta \qquad \sin(\pi + \theta) = -\sin\theta \qquad \sin(\theta \pm 2\pi) = \sin\theta$$

Exact Values:

angle \\ function	0	$\pi/6$	$\pi/4$	$\pi/3$	$\pi/2$	π	$3\pi/2$
sine	0	$\frac{1}{2}$	$\frac{1}{2}\sqrt{2}$	$\frac{1}{2}\sqrt{3}$	1	0	-1
cosine	1	$\frac{1}{2}\sqrt{3}$	$\frac{1}{2}\sqrt{2}$	$\frac{1}{2}$	0	-1	0
tangent	0	$\frac{1}{3}\sqrt{3}$	1	$\sqrt{3}$	*	0	*

EXAMPLES: Test the symmetry of the following polar-form curves.

1. $r = 2(1 - \cos\theta)$

Solution:

   x-axis: Substitute $(r, -\theta)$: $r = 2[1 - \cos(-\theta)]$
   $$= 2(1 - \cos\theta).$$

   Unchanged, so it is symmetric with respect to the *x*-axis.

   y-axis: Substitute $(r, \pi - \theta)$: $r = 2[1 - \cos(\pi - \theta)]$
   $$= 2(1 + \cos\theta).$$

   Changed; check other primary representation.

Substitute $(-r, 2\pi - \theta)$: $-r = 2[1 - \cos(2\pi - \theta)]$
$$= 2[1 - \cos(-\theta)]$$
$$= 2(1 - \cos\theta).$$

Changed, so it is not symmetric with respect to the y-axis.

origin: Substitute $(r, \theta + \pi)$: $r = 2[1 - \cos(\theta + \pi)]$
$$= 2(1 + \cos\theta).$$

Changed; check other primary representation.

Substitute $(-r, 2\pi + \theta)$: $-r = 2[1 - \cos(2\pi + \theta)]$
$$= 2(1 - \cos\theta)$$

Changed, so it is not symmetric with respect to the origin.

For this curve we have shown symmetry only with respect to the x-axis.

2. $r = 4\sin 2\theta$.

Solution:

x-axis: Substitute $(r, -\theta)$: $r = 4\sin 2(-\theta)$
$$= 4\sin(-2\theta)$$
$$= -4\sin 2\theta.$$

Changed; check other primary representation.

Substitute $(-r, \pi - \theta)$: $-r = 4\sin 2(\pi - \theta)$
$$= 4\sin(2\pi - 2\theta)$$
$$= 4\sin(-2\theta)$$
$$= -4\sin 2\theta;$$
$$r = 4\sin 2\theta.$$

Unchanged; symmetric with respect to the x-axis.

y-axis: Substitute $(r, \pi - \theta)$: $r = 4\sin 2(\pi - \theta)$
$$= 4\sin(2\pi - 2\theta)$$
$$= 4\sin(-2\theta)$$
$$= -4\sin 2\theta.$$

Changed; check other representation.

Substitute $(-r, 2\pi - \theta)$: $-r = 4\sin 2(2\pi - \theta)$
$$= 4\sin(4\pi - 2\theta)$$
$$= 4\sin(-2\theta)$$
$$= -4\sin 2\theta;$$
$$r = 4\sin 2\theta.$$

Unchanged; symmetric with respect to the y-axis.

Since the curve is symmetric with respect to both the x- and the y-axis, it must be symmetric with respect to the origin as well.

We now turn our attention to some polar-form graphing.

EXAMPLE: Graph $r = 2(1 - \cos \theta)$.

Solution: First check for symmetry. From Example 1 above this curve is symmetric with respect to the x-axis. Thus we check values of θ from 0 to π (first and second quadrants).

θ	0	$\pi/6$	$\pi/4$	$\pi/3$	$\pi/2$	$2\pi/3$	$3\pi/4$	$5\pi/6$	π
r	0	$2 - \sqrt{3}$	$2 - \sqrt{2}$	1	2	3	$2 + \sqrt{2}$	$2 + \sqrt{3}$	4

These points are plotted and then connected as shown in Figure 6.10.

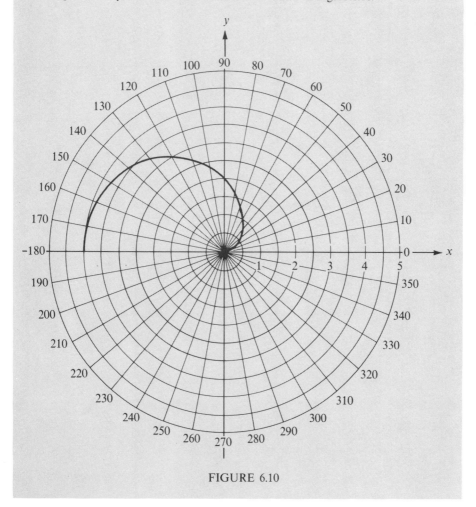

FIGURE 6.10

By symmetry, we plot the curve as shown in Figure 6.11. This curve is called a *cardioid*.

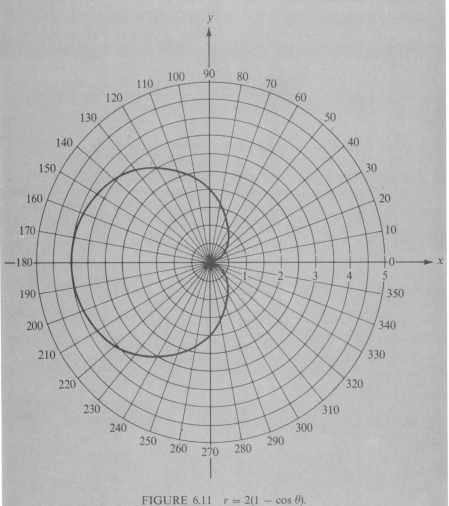

FIGURE 6.11 $r = 2(1 - \cos \theta)$.

EXAMPLE: Graph $r = 4 \sin 2\theta$.

Solution: From Example 2 above we know that this curve is symmetric with respect to the *x*-axis, the *y*-axis, and the origin. Thus we check values of θ from 0 to $\pi/2$.

θ	0	$\pi/12$	$\pi/8$	$\pi/6$	$\pi/4$	$\pi/3$	$3\pi/8$	$5\pi/12$	$\pi/2$
2θ	0	$\pi/6$	$\pi/4$	$\pi/3$	$\pi/2$	$2\pi/3$	$3\pi/4$	$5\pi/6$	π
r	0	2	$2\sqrt{2}$	$2\sqrt{3}$	4	$2\sqrt{3}$	$2\sqrt{2}$	2	0

These points—(r, θ) and *not* $(r, 2\theta)$—are plotted, and symmetry is used to sketch the curve, as shown in Figure 6.12. This curve is called a *four-leaved rose*.

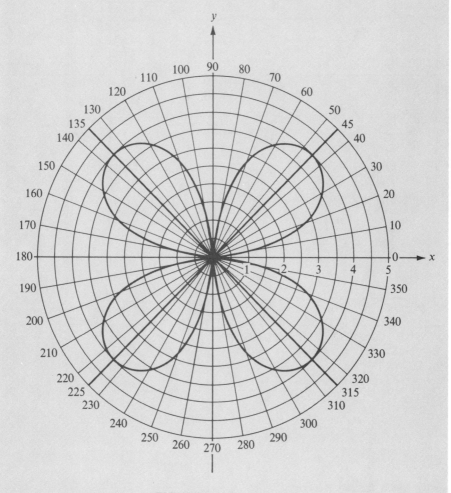

FIGURE 6.12 $r = 4 \sin 2\theta$.

Several special polar-form curves are described briefly in the remainder of this section. They would be graphed as shown in the above examples (the details are left for the reader). They are included here to make it easier for you to recognize the shape of a curve simply by looking at the equation.

Limacon: $r = a \pm b \cos \theta$ or $r = a \pm b \sin \theta$.

a. $a < b$ b. $a = b$; *cardioid*   c. $a > b$

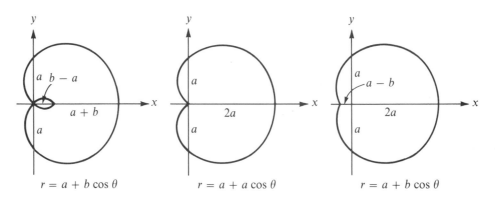

$r = a + b \cos \theta$ $r = a + a \cos \theta$ $r = a + b \cos \theta$

Lemniscate: $r^2 = a^2 \cos 2\theta$ or $r^2 = a^2 \sin 2\theta$.

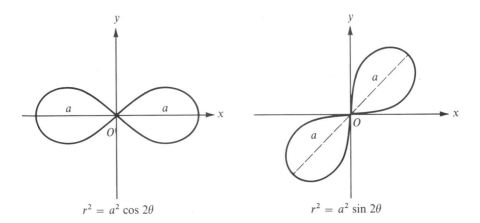

$r^2 = a^2 \cos 2\theta$ $r^2 = a^2 \sin 2\theta$

Rose Curves: $r = a \cos n\theta$ or $r = a \sin n\theta$ (*n* is a positive integer).

a. If *n* is odd, the rose is *n*-leaved.
b. If *n* is even, the rose is 2*n*-leaved.
c. If $n = 1$, the rose is a curve with one petal and is circular.

Two leaves (see lemniscate).

Three leaves:

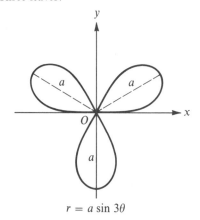

$$r = a \sin 3\theta$$

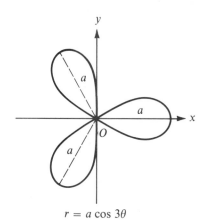

$$r = a \cos 3\theta$$

Four leaves:

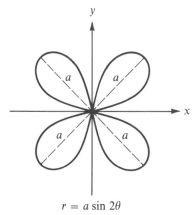

$$r = a \sin 2\theta$$

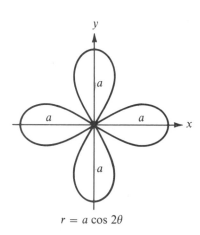

$$r = a \cos 2\theta$$

Spirals:

a. Spiral of Archimedes: $r = a\theta$.

b. Hyperbolic spiral: $r\theta = a$.

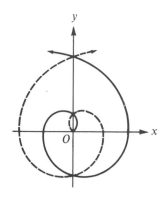

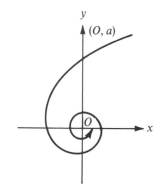

c. Logarithmic spiral: $r = a^{k\theta}$.

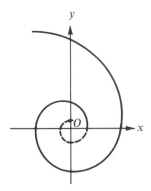

PROBLEM SET 6.6

A Problems

Identify each of the curves in Problems 1–12.

1. $r^2 = 9 \cos 2\theta$ 2. $r = 3\theta$ 3. $r = 4 \sin 30°$

4. $r = 2 \sin 2\theta$ 5. $r^2 = 2 \cos 2\theta$ 6. $r\theta = 3$

7. $r = 4$ 8. $r \cos \theta = -5$ 9. $r/(1 - \sin \theta) = 2$

10. $\theta = \tan \pi/4$ 11. $r = 2 - 3 \sin \theta$ 12. $\cos \theta = 1 - r$

13. Which of the curves that we've introduced is most descriptive of the path of the lawn mower in the following cartoon?

B Problems

In Problems 14–21 discuss symmetry and sketch the curve.

14. $r = 2(1 + \cos \theta)$ 15. $r = 2 \cos \theta$ 16. $r^2 = 4 \sin 2\theta$

17. $r = 6 \cos 3\theta$ 18. $r = \tan \theta$ 19. $r = 5 \sin 30°$

20. $r = 3\theta$ 21. $r = 2 \sin 3\theta$

22. Justify the rules for checking symmetry in polar form given on page 231.

23. Derive the equation for the distance between the points (r_1, θ_1) and (r_2, θ_2).

24. Find the equation of the circle with center $(3, \pi/6)$ and radius 5. (Hint: use the law of cosines.)

25. Derive the equation for the circle with center (c, α) and radius a.

Mind Boggler

26. Write an equation that will draw the heart in the *Peanuts* cartoon. The heart should be oriented and shaded as shown.

6.7 SUMMARY AND REVIEW

TERMS

Absolute value (6.2)
Argument (6.2)
Cardioid (6.6)
Complex conjugates (6.1)
Complex numbers (6.1)
Complex plane (6.2)
Conjugate (6.1)
De Moivre's Formula (6.4)
Gaussian plane (6.2)
i (6.1)
Imaginary axes (6.2)
Imaginary unit (6.1)
Lemniscate (6.6)
Limacon (6.6)

Modulus (6.2)
Parallelogram law (6.2)
Polar coordinate system (6.5)
Polar form of a complex number (6.2)
Pole (6.5)
Primary representations of a point (6.5)
Pure imaginary number (6.1)
r cis θ (6.2)
Real axes (6.2)
Rectangular form of a complex number (6.2)
Rose curve (6.6)
Spiral (6.6)
Symmetry (6.6)
Trigonometric form of a complex number (6.2)

CONCEPTS

(6.1)   The operations with complex numbers are carried out using the ordinary rules of algebra with the additional step of replacing i^2 by -1 whenever it occurs.

(6.2)   To change from the rectangular form of a complex number $a + bi$ to polar form, use:

$$r = \sqrt{a^2 + b^2};$$

$$\tan \theta = \frac{b}{a}.$$

(6.2)   To change from the polar form of a complex number r cis θ to rectangular form, use:

$$a = r \cos \theta;$$
$$b = r \sin \theta.$$

(6.3)   If r_1 cis θ_1 and r_2 cis θ_2 are complex numbers, then their product is

$$r_1 r_2 \operatorname{cis}(\theta_1 + \theta_2)$$

and their quotient is

$$\frac{r_1}{r_2} \operatorname{cis}(\theta_1 - \theta_2).$$

(6.4)   De Moivre's Formula: if n is any real number,

$$(r \text{ cis } \theta)^n = r^n \text{ cis } n(\theta + 360°k)$$

for any integer k.

(6.5)   To change from rectangular coordinates (x, y) to polar coordinates, use

$$r = \sqrt{x^2 + y^2};$$

$$\tan \theta = \frac{y}{x}.$$

(6.5)   To change from polar coordinates (r, θ) to rectangular coordinates, use

$$x = r \cos \theta;$$
$$y = r \sin \theta.$$

(6.5)   The primary representations of a point in polar coordinates are (r, θ) and $(-r, \pi + \theta)$. The other representations add multiples of 2π to the angles.

(6.6)   A polar-form equation is symmetric with respect to
1. the x-axis if the equation remains unchanged when we substitute

$$(r, -\theta) \quad \text{or} \quad (-r, \pi - \theta) \quad \text{for} \quad (r, \theta);$$

2. the y-axis if the equation remains unchanged when we substitute

$$(r, \pi - \theta) \quad \text{or} \quad (-r, 2\pi - \theta) \quad \text{for} \quad (r, \theta);$$

3. the origin if the equation remains unchanged when we substitute

$$(r, \pi + \theta) \quad \text{or} \quad (-r, 2\pi + \theta) \quad \text{for} \quad (r, \theta).$$

PROBLEM SET 6.7 *(Chapter 6 Test)*

1. (6.1) Simplify the expressions.
   a. $(2 - 5i) - (3 - 2i)$ b. $(2 - 5i)^2$
   c. $\dfrac{-4}{i}$ d. $\dfrac{2 + 3i}{1 - i}$

2. (6.2) Plot each pair of numbers, and show both the sum and the difference graphically.
   a. $4 + 2i, 1 + 5i$ b. $-4 - i, 4 - 2i$

3. (6.2) Write each of the following numbers in both rectangular and polar form. Also, graph each.
   a. $7 - 7i$ b. $-3i$
   c. $4(\cos 7\pi/4 + i \sin 7\pi/4)$ d. $2 \text{ cis } 150°$

4. (6.3) Perform the indicated operations, and leave your answers in rectangular form.
   a. $(\sqrt{12} - 2i)^4$ b. $\dfrac{(3 + 3i)(\sqrt{3} - i)}{1 + i}$

5. (6.3) Perform the indicated operations, and leave your answers in polar form.

   a. $2i(-1 + i)(-2 + 2i)$

   b. $\dfrac{2 \text{ cis } 158° \cdot 4 \text{ cis } 212°}{(2 \text{ cis } 312°)^3}$

6. (6.4) Find the indicated roots, and represent them in polar form, in rectangular form, and graphically.

   a. square roots of $\frac{7}{2}\sqrt{3} - \frac{7}{2}i$

   b. fourth roots of 1

7. (6.5) Give both primary representations of the following polar form points. Also, plot the points.

   a. $(5, \sqrt{75})$ b. $(3, -2\pi/3)$ c. $(-2, 2)$ d. $(-5, -\pi/10)$

8. (6.6) Identify each of the given curves.

   a. $r^2 = 9 \cos 2\theta$

   b. $r = 2 + 2 \sin \theta$

   c. $r = 4 \cos 5\theta$

   d. $r = 5 \sin \pi/4$

9. (6.6) Discuss the symmetry of and sketch $r = 2 + 2 \sin \theta$.

10. (6.6) Discuss the symmetry of and sketch $r = 4 \cos 5\theta$.

APPENDIX A

Selected Answers

PROBLEM SET 1.1 (PAGE 9)

1. Both **3.** Relation **5.** Relation **7.** Both **9.** Relation **11.** a. 1.07
b. 0.45 c. 0.21 d. 1.31 **13.** a. $0.20 b. $e(1974) - e(1944)$ **15.** a. 3 b. 0 c. 3
d. 12 **17.** a. -5 b. -7 c. -11 d. -15 **19.** a. 5 b. 25 c. 40 d. 50
21. a. w^2 b. h^2 c. $(w + h)^2$ d. $w^2 + h^2$ **23.** a. x^4 b. x c. $(x + h)^2$ d. x^2

25. a. $2xh + h^2$ b. $2x + h$ **27.** a. $\dfrac{17}{36}$ b. $\dfrac{2h^2 + 5h}{(h - 1)(h + 3)(h + 2)}$ **29.** a. 2

b. $\sqrt{b + 1}$ **31.** a. .018 b. .0125 c. .058 d. $\dfrac{s(1944 + h) - s(1944)}{h}$

33. $(1.4, 126.042)$, $(2.6, 184.158)$, $D: \{0 \le t \le 7.026875\}$

PROBLEM SET 1.2 (PAGE 20)

1. a. $\pi/6$ b. $\pi/2$ c. $3\pi/2$ d. $\pi/4$ e. 2π f. $\pi/3$ g. π
3.   **5.** 5 **7.** $7\sqrt{2}$ **9.** $2\sqrt{10}$

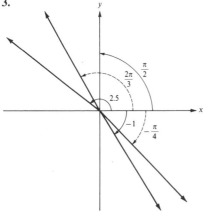

11. It is on the circle.
13. It is not on the circle.
15. It is on the circle.
17. It is on the circle.
19. $40°$ **21.** $30°$ **23.** π
25. $5\pi/3$ **27.** $330°$ **29.** $305°$ **31.** $7\pi/4$ **33.** $52.5°$ **35.** $146.83°$
37. $-127.17°$ **39.** $16.70°$ **41.** $14.514°$ **43.** $.207°$ **45.** $281.527°$
47. $\sqrt{2}$ **49.** $\sqrt{x^2 + [f(x)]^2}$ **51.** $\sqrt{(\beta - \alpha)^2 + (\omega - \theta)^2}$ **53.** $40°$
55. $6°$ **57.** $-165°$ **59.** $114.60°$ **61.** $-14.32°$ **63.** $22.92°$ **65.** $2\pi/9$
67. $-16\pi/45$ **69.** $127\pi/90$ **71.** 1.96 **73.** -1.10 **75.** 1 m
77. 31.40 m **79.** 70.69 cm **81.** 12.57 ft **83.** π cm or 3.14 cm **85.** 3979 mi
87. 890 km

PROBLEM SET 1.3 (PAGE 32)

1.

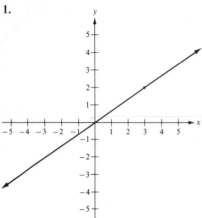

3.

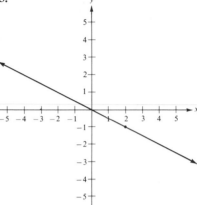

5.

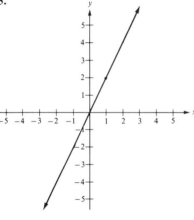

7.

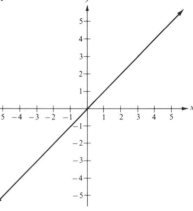

9.

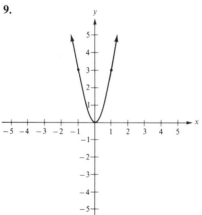

11.

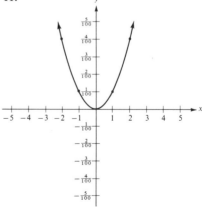

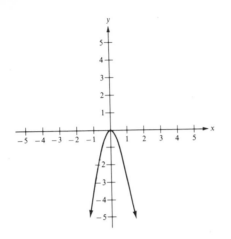

15.

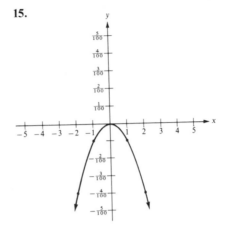

17.

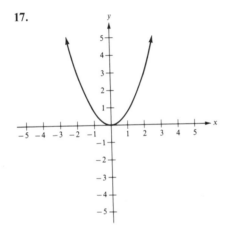

19.

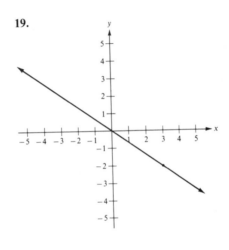

21.

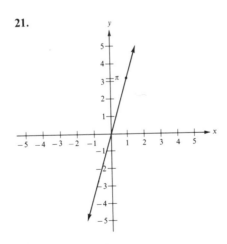

23.

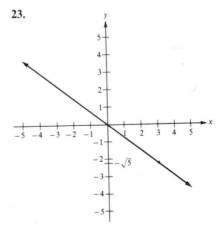

25.

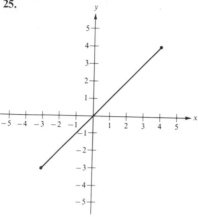

27.

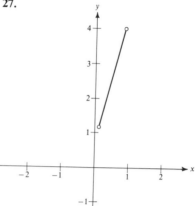

29.

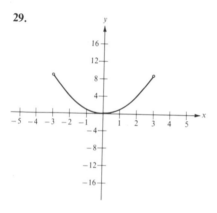

31.

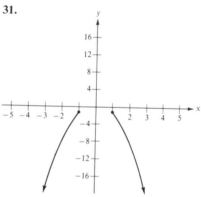

33.

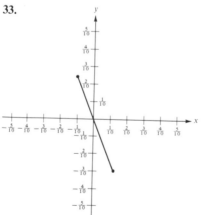

35.

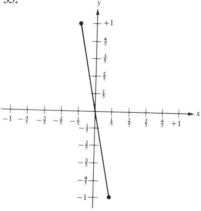

PROBLEM SET 1.4 (PAGE 40)

1.

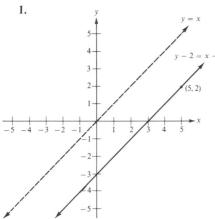

3.

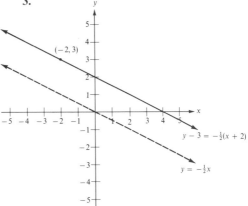

5.

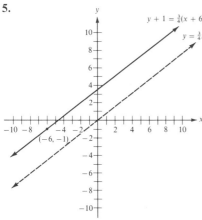

7.

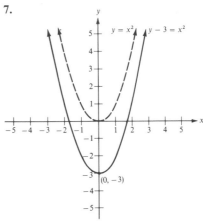

9.

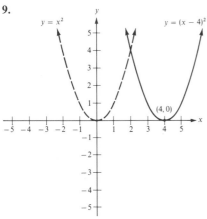

11.

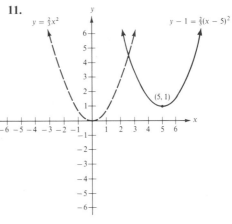

13.

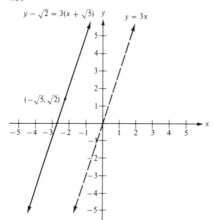

15.

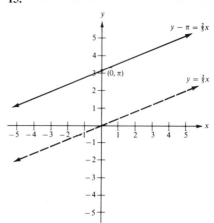

17.

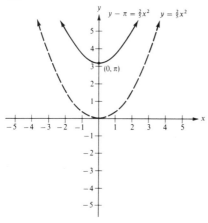

19.

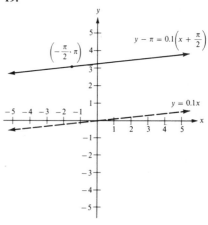

33. a. even b. even c. not even d. not even e. not even f. even

PROBLEM SET 1.5 (PAGE 50)

9. Yes   **11.** No   **13.** No   **15.** Yes   **17.** No

19. a. $f^{-1} = \{(5, 4), (3, 6), (1, 7), (4, 2)\}$ c. The inverse is a function. d. no change

21. a. $f^{-1} = \{(x, y) \mid y = \frac{1}{2}x - \frac{3}{2}\}$ c. The inverse is a function. d. no change

23. a. $f^{-1}(x) = x - 5$ c. The inverse is a function. d. no change

25. a. $y = \pm\sqrt{-2x}$ c. The inverse is not a function. d. $F(x) = -\frac{1}{2}x^2$ where $x \geq 0$.

Then $F^{-1}(x) = \sqrt{-2x}$, where $x \leq 0$.   **27.** a. $x = y$ if $y \geq 0$; $x = -y$ if $y < 0$

c. The inverse is not a function. d. $F = f$ where $x \geq 0$.   **29.** a. 1 b. 3 c. 6 d. -4

e. -2 **31.** a. 0 b. 2 c. 6 d. 7 e. 4   **33.** a. 0 b. 4 c. 0 d. 5

35. a. 0 b. 0 c. -3.5 d. 4

PROBLEM SET 1.6 (PAGE 58)

1. a. 9   b. 3   c. $4w - 3$ d. w e. x **3.** $6\sqrt{2}$ **5.** 50π m or about 157 m
7. a.

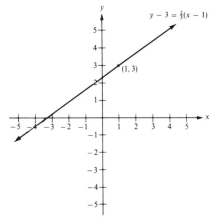

b.

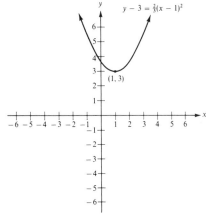

c.

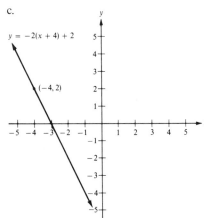

d.

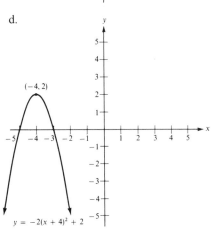

9. a. (a) $y = -\frac{2}{3}x$ (b) $y = -\frac{2}{3}x + 2$
(c) $y = \pm\frac{1}{3}\sqrt{-6x}\ (x \le 0)$
(d) $y = \pm\frac{1}{3}\sqrt{18 - 6x}\ (x \le 3)$
b. (a) and (b) are functions
(c) $y = \frac{1}{3}\sqrt{-6x}\ (x \ge 0)$ (d) $y = \frac{1}{3}\sqrt{18 - 6x}$
$(x \ge 0)$

PROBLEM SET 2.1 (PAGE 66)

1. $60°$ **3.** $\pi/4$ **5.** $\pi/2$ **7.** 0   **9.** $\sqrt{2}/2$ **11.** 1   **13.** -1
15. $\sqrt{3}/2$ **17.** 0   **19.** 0   **21.** $\frac{1}{2}$ **23.** $\sqrt{2}/2$ **25.** 0   **27.** $|x|\sqrt{2}$
29. $-x\sqrt{2}$ **31.** $\frac{1}{2}$ **33.** 0   **35.** 1   **37.** 1   **39.** $\frac{3}{4}$ **41.** $\frac{1}{4}$

PROBLEM SET 2.2 (PAGE 73)

1. a. 1 b. 1 c. $\frac{1}{2}\sqrt{3}$ d. $\frac{2}{3}\sqrt{3}$ e. undefined f. $\frac{1}{3}\sqrt{3}$ **3.** a. 1 b. 1 c. 0
d. $\frac{2}{3}\sqrt{3}$ e. $\frac{1}{2}\sqrt{2}$ f. 0   **5.** a. 1 b. -1 c. -1 d. 0 e. undefined f. undefined
7. 1   **9.** 1   **11.** a. $\sqrt{2}/2$ b. $\sqrt{2}/2$ **13.** a. 2 b. $\frac{1}{3}\sqrt{3}$ **15.** a. $-\sqrt{3}$
b. $-\sqrt{3}$ **17.** $\cos\theta = 3/5$; $\sin\theta = 4/5$; $\tan\theta = 4/3$; $\sec\theta = 5/3$; $\csc\theta = 5/4$;
$\cot\theta = 3/4$ **19.** $\cos\theta = -3/5$; $\sin\theta = -4/5$; $\tan\theta = 4/3$; $\sec\theta = -5/3$;
$\csc\theta = -5/4$; $\cot\theta = 3/4$ **21.** $\cos\theta = 5/13$; $\sin\theta = 12/13$; $\tan\theta = 12/5$; $\sec\theta = 13/5$;
$\csc\theta = 13/12$; $\cot\theta = 5/12$ **23.** $\cos\theta = 5/13$; $\sin\theta = -12/13$; $\tan\theta = -12/5$;
$\sec\theta = 13/5$; $\csc\theta = -13/12$; $\cot\theta = -5/12$ **25.** $\cos\theta = 2/\sqrt{29}$; $\sin\theta = -5/\sqrt{29}$;
$\tan\theta = -5/2$; $\sec\theta = \sqrt{29}/2$; $\csc\theta = -\sqrt{29}/5$; $\cot\theta = -2/5$ **27.** $\cos\theta = -4/\sqrt{41}$;
$\sin\theta = -5/\sqrt{41}$; $\tan\theta = 5/4$; $\sec\theta = -\sqrt{41}/4$; $\csc\theta = -\sqrt{41}/5$; $\cot\theta = 4/5$
29. $\cos 450° = 0$; $\sin 450° = 1$; $\tan 450°$ undefined; $\sec 450°$ undefined; $\csc 450° = 1$;
$\cot 450° = 0$ **31.** $\cos(-\pi/6) = \sqrt{3}/2$; $\sin(-\pi/6) = -\frac{1}{2}$; $\tan(-\pi/6) = -\sqrt{3}/3$;
$\sec(-\pi/6) = 2/\sqrt{3}$ or $(2/3)\sqrt{3}$; $\csc(-\pi/6) = -2/\sqrt{3}$ or $-(2/3)\sqrt{3}$; $\cot(-\pi/6) = -3/\sqrt{3}$
or $-\sqrt{3}$ **33.** $\cos(-2\pi) = 1$; $\sin(-2\pi) = 0$; $\tan(-2\pi) = 0$; $\sec(-2\pi) = 1$; $\csc(-2\pi)$
and $\cot(-2\pi)$ undefined   **35.** $\cos 390° = \sqrt{3}/2$; $\sin 390° = 1/2$; $\tan 390° = \sqrt{3}/3$;
$\sec 390° = 2/\sqrt{3}$; $\csc 390° = 2$; $\cot 390° = 3/\sqrt{3}$ or $\sqrt{3}$ **37.** $\cos(-135°) = -\sqrt{2}/2$;
$\sin(-135°) = -\sqrt{2}/2$; $\tan(-135°) = 1$; $\sec(-135°) = -2/\sqrt{2}$ or $-\sqrt{2}$; $\csc(-135°)$
$= -2/\sqrt{2}$ or $-\sqrt{2}$; $\cot(-135°) = 1$ **39.** $\cos 3\pi = -1$; $\sin 3\pi = 0$; $\tan 3\pi = 0$;
$\sec 3\pi = -1$; $\csc 3\pi$ and $\cot 3\pi$ undefined   **41.** 0.6   **43.** 2.0   **45.** -0.4
47. -1.0 **49.** -0.6 **51.** 0.2   **53.** $(47) - 1.015427$ $(48)\,0.342020$ $(49)\,-0.642788$
$(50)\,5.671282$ $(51)\,0.176327$ **55.** $(47) -1.0154$ $(48)\,0.3420$ $(49)\,-0.6428$ $(50)\,5.6713$
$(51)\,0.1763$ **57.** 0.205271   **59.** 0.841471   **61.** 0.886995   **63.** 0.366176

PROBLEM SET 2.3 (PAGE 84)

1. a. undefined b. 0 c. 0 d. $\sqrt{3}/2$ e. undefined f. 0   **3.** a. undefined b. -1
c. 0 d. 1 e. 1 f. 1/2   **5.** a. 1 b. $\sqrt{2}$ c. $\sqrt{3}$ d. 1 e. 0 f. -1

19.

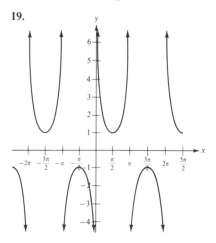

21.

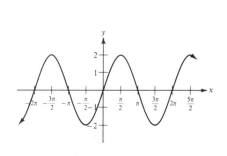

23.

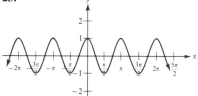

25.

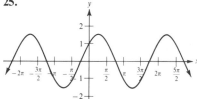

27.

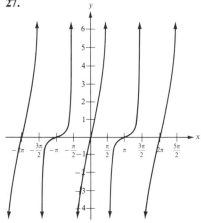

29.

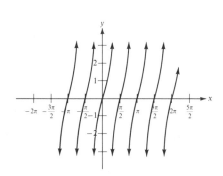

PROBLEM SET 2.4 (PAGE 94)

1. a. 1 b. undefined c. 0 d. undefined e. -1 f. 0   **3.** a. $\sqrt{3}/3$ b. 2
c. $\sqrt{2}/2$ d. $\sqrt{2}/2$ e. $\sqrt{3}$ f. undefined   **5.** a. $\sqrt{3}$ b. 0 c. $2\sqrt{3}/3$ d. $\sqrt{2}$
e. -1 f. $\sqrt{3}/3$

7.

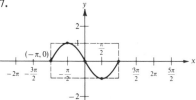

9.

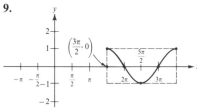

11.

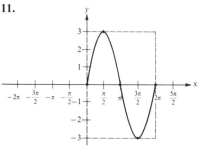

13.

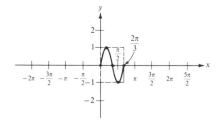

15.

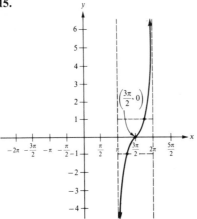

17.

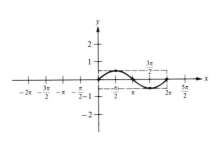

19.

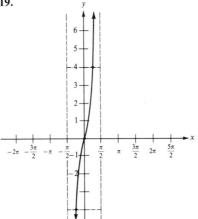

21.

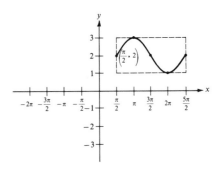

23.

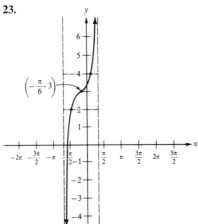

25.

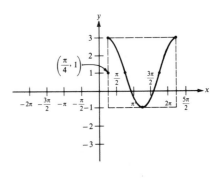

27.

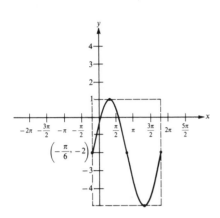

29.

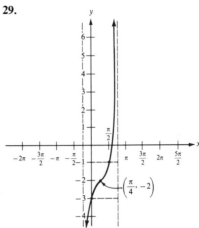

31.

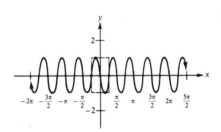

33.

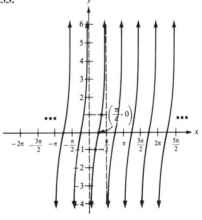

35.

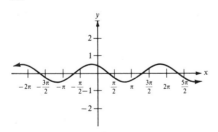

37.

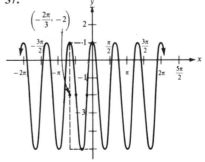

39.

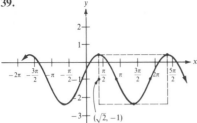

41.

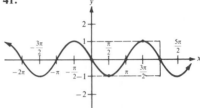

43.

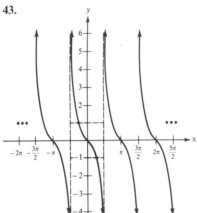

45.

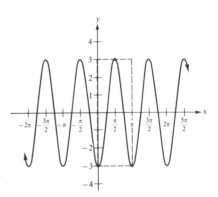

47.

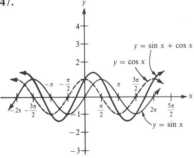

49.

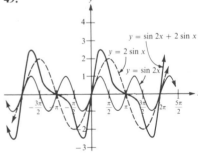

51.

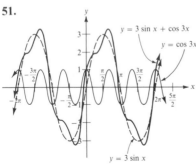

PROBLEM SET 2.5 (PAGE 104)

1. a. 0 b. 30° c. 30° d. 0   **3.** a. 60° b. 45° c. 45° d. 45°   **5.** a. 150°
b. −45° c. −45° d. 120°   **7.** a. 0 b. 60° c. 60° d. 60°   **9.** 33°   **11.** 85°
13. 19°   **15.** 54°   **17.** 108°   **19.** 69°   **21.** −66°   **23.** 1   **25.** $\frac{1}{3}$
27. $2\pi/15$ **29.** 0.4163
Note: In Problems 31–39 n represents any integer.

31. $\begin{cases} 30° \pm 360°n \\ 150° \pm 360°n \end{cases}$ **33.** $60° \pm 180°n$ **35.** $\begin{cases} 150° \pm 360°n \\ 210° \pm 360°n \end{cases}$ **37.** $\begin{cases} 23° \pm 360°n \\ 157° \pm 360°n \end{cases}$

39. $54° \pm 180°n$

41.

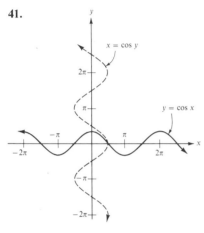

43.

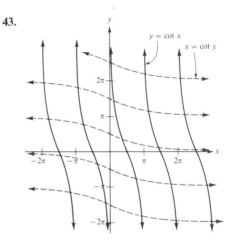

45.

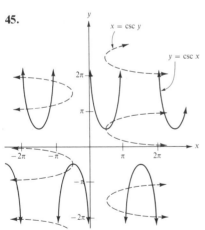

PROBLEM SET 2.6 (PAGE 107)

1. a. 1/2 b. 1/2 c. 1 d. 1 e. undefined f. $\sqrt{3}/3$ g. $\sqrt{2}/2$ h. $\sqrt{3}$ **3.** a. 1/2
b. 1/2 c. 1   **5.** a. 1.089615865 b. −1.338648128 c. 1.458652446 d. 18.3°
or 18° 20′ e. 130.5° or 130° 30′ f. 73°

7. a. $y - 2 = \sin(x - \pi/6)$;
    $(h, k) = (\pi/6, 2)$; amp $= 1$;
    period $= 2\pi$

b. $y = \cos(x + \pi/4)$;
    $(h, k) = (-\pi/4, 0)$; amp $= 1$;
    period $= 2\pi$

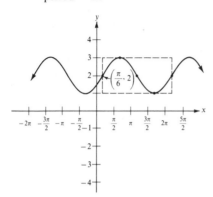

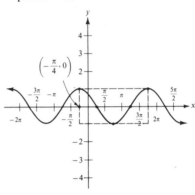

c. $y = \tan(x - \pi/3) - 2$;
    $(h, k) = (\pi/3, -2)$; amp $= 1$;
    period $= \pi$

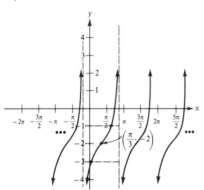

9. a. $y = 2\cos(2x + \pi/3)$;
    $y = 2\cos 2(x + \pi/6)$; $(h,k) = (-\pi/6, 0)$;
    amp $= 2$; period $= \pi$

b. $y = 2\sin 2x + \cos x$

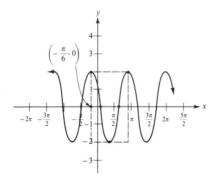

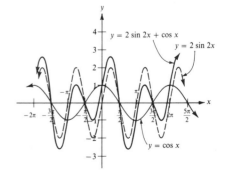

PROBLEM SET 3.1 (PAGE 114)

3. III, IV   **5.** I, IV   **7.** II   **9.** I   **11.** $\cot(A + B)$ **13.** $\tan(\pi/15)$
15. $\cos(\pi/8)$ **17.** $\cos 127°$ **19.** $\sec^2(\pi/5)$ **21.** 1   **23.** -1 **25.** 1
31. $\tan \theta = \tan \theta$; $\cot \theta = 1/\tan \theta$; $\sec \theta = \pm\sqrt{1 + \tan^2 \theta}$; $\cos \theta = 1/\pm\sqrt{1 + \tan^2 \theta}$;
$\sin \theta = \tan \theta/\pm\sqrt{1 + \tan^2 \theta}$; $\csc \theta = \pm\sqrt{1 + \tan^2 \theta}/\tan \theta$ **33.** $\sec \theta = \sec \theta$;
$\cos \theta = 1/\sec \theta$; $\tan \theta = \pm\sqrt{\sec^2 \theta - 1}$; $\cot \theta = 1/\pm\sqrt{\sec^2 \theta - 1}$;
$\sin \theta = \pm\sqrt{\sec^2 \theta - 1}/\sec \theta$; $\csc \theta = \sec \theta/\pm\sqrt{\sec^2 \theta - 1}$ **35.** $\sin \theta = -12/13$;
$\tan \theta = -12/5$; $\sec \theta = 13/5$; $\csc \theta = -13/12$; $\cot \theta = -5/12$ **37.** $\cot \theta = 12/5$;
$\sec \theta = 13/12$; $\cos \theta = 12/13$; $\sin \theta = 5/13$; $\csc \theta = 13/5$ **39.** $\cos \theta = \sqrt{5}/3$;
$\tan \theta = 2\sqrt{5}/5$; $\cot \theta = \sqrt{5}/2$; $\csc \theta = 3/2$; $\sec \theta = 3\sqrt{5}/5$ **41.** $\cos \theta = 5\sqrt{34}/34$;
$\tan \theta = -3/5$; $\cot \theta = -5/3$; $\sin \theta = -3\sqrt{34}/34$; $\csc \theta = -\sqrt{34}/3$
43. $\sin \theta = -3\sqrt{10}/10$; $\cos \theta = \sqrt{10}/10$; $\cot \theta = -1/3$; $\tan \theta = -3$; $\sec \theta = \sqrt{10}$

PROBLEM SET 3.4 (PAGE 131)

1. $\sin(-15°) = (\sqrt{2} - \sqrt{6})/4$; $\cos(-15°) = (\sqrt{6} + \sqrt{2})/4$; $\tan(-15°) = -2 + \sqrt{3}$
3. $\sin 75° = (\sqrt{2} + \sqrt{6})/4$; $\cos 75° = (\sqrt{6} - \sqrt{2})/4$; $\tan 75° = 2 + \sqrt{3}$
5. $\sin 345° = (\sqrt{2} - \sqrt{6})/4$; $\cos 345° = (\sqrt{6} + \sqrt{2})/4$; $\tan 345° = -2 + \sqrt{3}$
7. $\sin 225° = -\sqrt{2}/2$; $\cos 225° = -\sqrt{2}/2$; $\tan 225° = 1$
9. $\sin 285° = (-\sqrt{2} - \sqrt{6})/4$; $\cos 285° = (\sqrt{6} - \sqrt{2})/4$; $\tan 285° = -2 - \sqrt{3}$
11. $(\sqrt{3}\cos \theta - \sin \theta)/2$ **13.** $(1 + \tan \theta)/(1 - \tan \theta)$ **15.** $(\sqrt{3}\cos \theta - \sin \theta)/2$
17. $\cos^2 \theta - \sin^2 \theta$ **19.** $(2\tan \theta)/(1 - \tan^2 \theta)$ **21.** 0.8746   **23.** 0.6561
25. 0.6745   **31.** 63/65   **33.** 56/33   **35.** $-455/697$ **37.** $-455/528$

PROBLEM SET 3.5 (PAGE 139)

1. $\sqrt{2}/2$ **3.** $\frac{1}{2}$ **5.** -1 **7.** $\frac{1}{2}\sqrt{2 - \sqrt{2}}$ **9.** $\sqrt{2} - 1$
11. $\cos \frac{1}{2}\theta = 3\sqrt{10}/10$; $\sin \frac{1}{2}\theta = \sqrt{10}/10$; $\tan \frac{1}{2}\theta = 1/3$; $\cos 2\theta = 7/25$; $\sin 2\theta = 24/25$;
$\tan 2\theta = 24/7$ **13.** $\cos \frac{1}{2}\theta = -5\sqrt{26}/26$; $\sin \frac{1}{2}\theta = \sqrt{26}/26$; $\tan \frac{1}{2}\theta = -1/5$;
$\cos 2\theta = 119/169$; $\sin 2\theta = -120/169$; $\tan 2\theta = -120/119$ **15.** $\cos \frac{1}{2}\theta = \sqrt{7}/3$;
$\sin \frac{1}{2}\theta = \sqrt{2}/3$; $\tan \frac{1}{2}\theta = \sqrt{14}/7$; $\cos 2\theta = -31/81$; $\sin 2\theta = 20\sqrt{14}/81$;
$\tan 2\theta = -20\sqrt{14}/31$ **17.** $\cos 40° + \cos 110°$ **19.** $\cos 11° - \cos 59°$
21. $\frac{1}{2}\cos 18° - \frac{1}{2}\cos 158°$ **23.** $\frac{1}{2}\sin 60° + \frac{1}{2}\sin 22°$ **25.** $\frac{1}{2}\cos 2\theta + \frac{1}{2}\cos 4\theta$
27. $2\sin 8° \cos 14°$ **29.** $2\cos 51.5° \cos 26.5°$ **31.** $2\sin(3x/2)\cos(x/2)$
33. $2\cos(5\theta/2)\cos(\theta/2)$ **35.** $\cos \theta = \sqrt{2}/2$; $\sin \theta = \sqrt{2}/2$; $\tan \theta = 1$
37. $\cos \theta = -1/2$; $\sin \theta = \sqrt{3}/2$; $\tan \theta = \sqrt{3}$
39. $\cos \theta = 3\sqrt{10}/10$; $\sin \theta = \sqrt{10}/10$; $\tan \theta = 1/3$

PROBLEM SET 3.6 (PAGE 143)

1. 60°, 300°   **3.** 22.5°, 67.5°, 202.5°, 247.5°   **5.** 15°, 75°, 135°, 195°, 255°, 315°
7. 0°, 90°, 180°, 270°   **9.** 60°, 90°, 300°   **11.** 0°, 60°, 180°, 240°
13. 0°, 70.53°, 180°, 289.47°   **15.** 30°, 150°, 210°, 330°   **17.** 45°, 225°
19. 45°, 135°, 225°, 315°   **21.** 270°   **23.** 60°, 300°   **25.** 51.83°, 308.17°
27. 16.31°, 163.69°   **29.** 30°, 60°, 210°, 240°

PROBLEM SET 3.7 (PAGE 146)

3. $\cos \theta = -4/5$; $\tan \theta = -3/4$; $\csc \theta = 5/3$; $\sec \theta = -5/4$; $\cot \theta = -4/3$ **9.** $270°$

PROBLEM SET 4.1 (PAGE 152)

Answers may vary.   **1.** $A = 90° - B$; $b = 80 \tan 60°$; $c = 80/\cos 60°$
3. $A = \tan^{-1}(68/83)$; $B = 90° - A$; $c = 83/\cos A$ **5.** $A = 90° - B$; $a = 13/\tan 65°$;
$c = 13/\sin 65°$ **7.** $A = 90° - B$; $a = 90 \tan A$; $c = 90/\sin 13°$
9. $A = \tan^{-1}(29/24)$; $B = 90° - A$; $c = 24/\cos A$ **11.** $B = 90° - A$;
$a = 28.3 \sin 69.2°$; $b = 28.3 \cos 69.2°$ **13.** $A = 90° - B$; $b = 70 \tan 57.4°$;
$c = 70/\cos 57.4°$ **15.** $A = 90° - B$; $b = 9000 \tan 23°$; $c = 9000/\cos 23°$
17. $A = 90° - B$; $a = 2580/\tan 16.4°$; $c = 2580/\sin 16.4°$ **19.** $A = \cos^{-1}(3200/7700)$;
$B = 90° - A$; $a = 7700 \sin A$

PROBLEM SET 4.2 (PAGE 156)

1. $A = 30°$; $B = 60°$; $C = 90°$; $a = 80$; $b = 140$; $c = 160$
3. $A = 39°$; $B = 51°$; $C = 90°$; $a = 68$; $b = 83$; $c = 110$
5. $A = 25°$; $B = 65°$; $C = 90°$; $a = 6.1$; $b = 13$; $c = 14$
7. $A = 77°$; $B = 13°$; $C = 90°$; $a = 390$; $b = 90$; $c = 400$
9. $A = 50°$; $B = 40°$; $C = 90°$; $a = 24$; $b = 29$; $c = 45$
11. $A = 69.2°$; $B = 20.8°$; $C = 90.0°$; $a = 26.5$; $b = 10.0$; $c = 28.3$
13. $A = 32.6°$; $B = 57.4°$; $C = 90.0°$; $a = 70.0$; $b = 109$; $c = 130$
15. $A = 67°$; $B = 23°$; $C = 90°$; $a = 9000$; $b = 3800$; $c = 9800$
17. $A = 73.6°$; $B = 16.4°$; $C = 90°$; $a = 8780$; $b = 2580$; $c = 9140$
19. $A = 65.44°$; $B = 24.56°$; $C = 90.00°$; $a = 7003$; $b = 3200$; $c = 7700$
21. 36.15 **23.** 79.3 **25.** 5.25 **27.** 59.4 **29.** 12.2
31. 0.0656 **33.** 2.26 **35.** 0.263 **37.** 3.535 **39.** $33°$
41. $A = 67.8°$; $B = 22.2°$; $C = 90.0°$; $a = 26.6$; $b = 10.9$; $c = 28.7$
43. $A = 28.95°$; $B = 61.05°$; $C = 90.00°$; $a = 202.7$; $b = 366.4$; $c = 418.7$
45. $A = 48.3°$; $B = 41.7°$; $C = 90.0°$; $a = 0.355$; $b = 0.316$; $c = 0.475$
47. $A = 13.94°$; $B = 76.06°$; $C = 90.00°$; $a = 0.0893$; $b = 0.3596$; $c = 0.3705$
49. $A = 61.91°$; $B = 28.09°$; $C = 90.00°$; $a = 64.70$; $b = 34.53$; $c = 73.33$
51. $A = 59.2°$; $B = 30.8°$; $C = 90.0°$; $a = 160$; $b = 95.2$; $c = 186$
53. $A = 11.9°$; $B = 78.1°$; $C = 90.0°$; $a = 8.85$; $b = 42.0$; $c = 42.8$
55. $58°$ with the 28-m side; $32°$ with the 45-m side.

PROBLEM SET 4.3 (PAGE 161)

1. The building is 23 m tall.   **3.** The ship is 200 m away.   **5.** The distance is
350 ft.   **7.** The top of the ladder is 13 ft above the ground.   **9.** The angle of
elevation is about 56°.   **11.** The height is 1251 ft.   **13.** The distance is 780 ft.
15. The height of the Empire State Building is 1250 ft, and that of the Sears Tower is
1454 ft.   **17.** The distance is 51.8 ft.   **19.** The center is 14.7 ft high.
21. The plate should be placed at 12 ft for a pitch of 5/12.   **23.** The distance from
the earth to Venus is 63,400,000 (or 6.34×10^7) mi.   **25.** The radius of the inscribed
circle is 633.9 ft.   **29.** The shadow will be 12 ft long.

PROBLEM SET 4.4 (PAGE 170)

1. 5.0; 53°   **3.** 23; 52°   **5.** 22; 7.1   **7.** The boat is traveling at 13 mph with a bearing of S22°W.   **9.** A force of 4.71 tons is needed to keep it from rolling down the hill. The force against the hill is 51.8 tons.   **11.** In one hour it has traveled 590 miles south.   **13.** The horizontal component is 2090 fps, and the vertical component is 368 fps.   **15.** The angle of inclination is 39.2°.   **17.** The pilot must fly in the direction of 4.9° and the speed relative to the ground is 240 mph.   **19.** The horizontal component is 3.19 ft/sec², and the vertical component is 2.90 ft/sec².   **21.** The heaviest piece of cargo is 12,200 pounds.   **23.** The object is projected S77°E with a speed of 870 m per minute.

PROBLEM SET 4.5 (PAGE 173)

1. a. $b \tan A$ b. $c \sin A$ c. $b/\tan B$ d. $c \cos B$ **3.** a. $\sqrt{a^2 + b^2}$ b. $a/\sin A$ c. $b/\cos A$ d. $a/\cos B$ **5.** a. $\tan^{-1}(b/a)$ b. $\cos^{-1}(a/c)$ c. $\sin^{-1}(b/c)$ d. $90° - A$ **7.** The tower is 2063 ft high.   **9.** The falls are 2643 ft high.

PROBLEM SET 5.1 (PAGE 182)

1. $\alpha = 54°$ **3.** $\gamma = 49°$ **5.** The largest angle is $\alpha = 80°$.   **7.** $c = 10$ **9.** $a = 17$ **15.** The largest angle is $\beta = 78°$.   **17.** $\beta = 46.6°$ **19.** $a = 604$

PROBLEM SET 5.2 (PAGE 185)

1. $\alpha = 48°; \beta = 62°; \gamma = 70°; a = 10; b = 12; c = 13$
3. $\alpha = 30°; \beta = 50°; \gamma = 100°; a = 30; b = 46; c = 59$
5. $\alpha = 50°; \beta = 70°; \gamma = 60°; a = 33; b = 40; c = 37$
7. $\alpha = 82°; \beta = 19°; \gamma = 79°; a = 53; b = 18; c = 53$
9. $\alpha = 35.0°; \beta = 81.0°; \gamma = 64.0°; a = 73.4; b = 126; c = 115$
19. $\alpha = 98.0°; \beta = 28.7°; \gamma = 53.3°; a = 114; b = 55.0; c = 92.0$
21. $\alpha = 85.2°; \beta = 38.7°; \gamma = 56.1°; a = 148; b = 92.7; c = 123$
23. Cannot be solved unless at least one side is given.
25. Cannot be solved, since the sum of the two smaller sides is not larger than the third side.
27. $\alpha = 85.26°; \beta = 24.45°; \gamma = 70.29°; a = 2971; b = 1234; c = 2807$

PROBLEM SET 5.3 (PAGE 192)

1. $\alpha > 90°$ and $a < b$; no triangle formed.   **3.** No triangle formed.   **5.** $\alpha = 125°$; $\beta = 41°; \gamma = 14°; a = 5.0; b = 4.0; c = 1.5$ **7.** $\alpha = 75°; \beta = 49°; \gamma = 56°; a = 9.0;$ $b = 7.0; c = 7.7$ **9.** $a < h$; no triangle formed.   **11.** $\alpha = 47.0°; \beta = 90.0°;$ $\gamma = 43.0°; a = 8.63; b = 11.8; c = 8.05$ **13.** $a < h$; no triangle formed.
15. $\alpha = 56°; \beta = 36°; \gamma = 88°; a = 7.0; b = 5.0; c = 8.4$
17. $\alpha = 20.9°; \beta = 24.1°; \gamma = 135.0°; a = 14.2; b = 16.3; c = 28.2$
19. $\alpha = 82.3°; \beta = 61.5°; \gamma = 36.2°; a = 51.5; b = 45.7; c = 30.7$
21. No solution, since the sum of the two smaller sides must be larger than the third side.

23. No solution, since at least one side must be given.   **25.** Solution I:
$\alpha = 43.7°$; $\beta = 71.2°$; $\gamma = 65.1°$; $a = 27.2$; $b = 37.3$; $c = 35.7$.   Solution II: $\alpha = 43.7°$;
$\beta' = 21.4°$; $\gamma' = 114.9°$; $a = 27.2$; $b' = 14.4$; $c = 35.7$ **27.** Solution I: $\alpha = 120.6°$;
$\beta = 31.0°$; $\gamma = 28.4°$; $a = 715$; $b = 428$; $c = 395$.   Solution II: $\alpha' = 2.6°$; $\beta = 149.0°$;
$\gamma = 28.4°$; $a' = 37.7$; $b = 428$; $c = 395$ **29.** $\alpha = 31.4°$; $\beta = 12.4°$; $\gamma = 136.2°$; $a = 612$;
$b = 251$; $c = 812$

PROBLEM SET 5.4 (PAGE 196)

1. The lengths of the diagonals are 18 m and 39 m.   **3.** He is 5.4 mi from the target.
5. The height of the tree is 58 ft.   **7.** The direction is S32.8°W with a speed of 22.6
knots.   **9.** The distance from L to T is 454 ft.   **11.** The altitude of the UFO is
1850 ft (.35 mi). The distance is 1.926 mi (10,170 ft) from the first city and .5363 mi
(2832 ft) from the second city.   **13.** The angle of elevation is 19.0°.   **15.** The jetty
is about 5030 ft long.   **17.** The boats are 67.2 mi apart.   **19.** About 5.07 acres
will cost about $127.   **25.** Area is 180 sq units.   **27.** Area is 1140 sq units.
29. Area is 20,100 sq units.

PROBLEM SET 5.5 (PAGE 200)

1. Procedure: $\cos \alpha = \dfrac{b^2 + c^2 - a^2}{2bc}$.   Solution: $\alpha = \cos^{-1}\left(\dfrac{b^2 + c^2 - a^2}{2bc}\right)$

3. Procedure: $\alpha + \beta + \gamma = 180°$ Solution: $\alpha = 180° - \beta - \gamma$
5. Procedure: $b^2 = a^2 + c^2 - 2ac \cos \beta$.   Solution: $b = \sqrt{a^2 + c^2 - 2ac \cos \beta}$
7. Let h be an altitude drawn from B. Then $\sin \alpha = h/c$ and $\sin \gamma = h/a$.
Thus, $c \sin \alpha = a \sin \gamma$ and $\sin \alpha/a = \sin \gamma/c$.   **9.** 113 m of fence is needed.

PROBLEM SET 6.1 (PAGE 207)

1. $8 + 7i$ **3.** $1 - 6i$ **5.** -5 **7.** -4 **9.** $11 - i$ **11.** $-3 + 2i$
13. $7 + i$ **15.** 29 **17.** 34 **19.** 1 **21.** $-i$ **23.** -1 **25.** 1
27. $-i$ **29.** -1 **31.** $32 - 24i$ **33.** $-9 + 40i$.   **35.** $-\frac{3}{2} + \frac{3}{2}i$
37. $-2i$ **39.** $-\frac{6}{29} + \frac{15}{29}i$ **41.** $1 + i$ **43.** $-\frac{35}{37} - \frac{12}{37}i$

PROBLEM SET 6.2 (PAGE 213)

1.

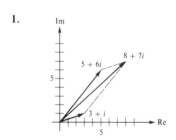

3.

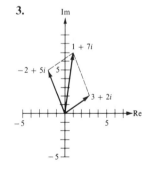

5.

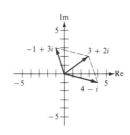

7.

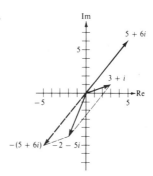

9.

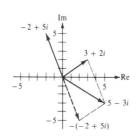

11.

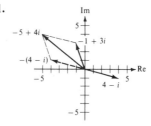

Note: The numbers in Problems 13–43 should also be graphed.   **13.** $\sqrt{2}$ cis 45°
15. 2 cis 330°   **17.** 2 cis 300°   **19.** 4 cis 330°   **21.** cis 0°   **23.** cis 270°
25. 3 cis 55°   **27.** 4 cis 100°   **29.** $\sqrt{2} + \sqrt{2}i$ **31.** $2\sqrt{2} - 2\sqrt{2}i$
33. $-\sqrt{3}/2 + \frac{1}{2}i$ **35.** 1   **37.** $-\frac{3}{2} + (3\sqrt{3}/2)i$ **39.** 4.2262 + 9.0631i
41. $-7\sqrt{2}/2 + (7\sqrt{2}/2)i$ **43.** $-8.8633 - 1.5628i$

PROBLEM SET 6.3 (PAGE 218)

1. 6 cis 210°   **3.** 48 cis 152°   **5.** 2 cis 15°   **7.** $\frac{5}{2}$ cis 267°   **9.** 2 cis 190°
11. 18 cis 167°   **13.** 8 cis 150°   **15.** cis 330°   **17.** 64 cis 180°   **19.** 8 cis 270°
21. $(\sqrt{2}/2)$ cis 75°   **23.** 4 cis 330°   **25.** $12\sqrt{2}$ cis 45°   **27.** $4\sqrt{2}$ cis 225°
29. cis 30°   Note: Problems 31–45 should also be graphed.   **31.** a. $6\sqrt{3} - 6i$
b. 12 cis 330°   **33.** a. $(3\sqrt{3} + 3) + (3\sqrt{3} - 3)i$ b. $6\sqrt{2}$ cis 15°   **35.** a. $-3 + 3\sqrt{3}i$
b. 6 cis 120°   **37.** a. $-3i$ b. 3 cis 270°   **39.** a. $(\frac{3}{2}\sqrt{3} - \frac{3}{2}) + (-\frac{3}{2}\sqrt{3} - \frac{3}{2})i$
b. $3\sqrt{2}$ cis 285°   **41.** a. $\frac{3}{4} + \frac{3}{4}\sqrt{3}i$ b. $\frac{3}{2}$ cis 60°   **43.** a. $-4.3079 + 34.3832i$
b. 36 cis 321°   **45.** a. 0.3131 + 2.228i b. $\frac{9}{4}$ cis 82°

PROBLEM SET 6.4 (PAGE 223)

1. 3 cis 120°, 3 cis 300°   **3.** 5 cis 150°, 5 cis 330°   **5.** 3 cis 100°, 3 cis 220°,
3 cis 340°   **7.** 2 cis 88°, 2 cis 178°, 2 cis 268°, 2 cis 358°   **9.** 4 cis 56°, 4 cis 146°,
4 cis 236°, 4 cis 326°   **11.** 2 cis 32°, 2 cis 104°, 2 cis 176°, 2 cis 248°, 2 cis 320°
13. cis 60°, cis 180°, cis 300°   **15.** 3 cis 0°, 3 cis 120°, 3 cis 240°   **17.** $\sqrt[8]{2}$ cis 56.25°,
$\sqrt[8]{2}$ cis 146.25°, $\sqrt[8]{2}$ cis 236.25°, $\sqrt[8]{2}$ cis 326.25°
19. cis 54°, cis 126°, cis 198°, cis 270°, cis 342°

Note: Problems 21–25 should also be illustrated graphically.
21. $1, -\frac{1}{2} + (\sqrt{3}/2)i, -\frac{1}{2} - (\sqrt{3}/2)i$ **23.** $1 + \sqrt{3}i, -2, 1 - \sqrt{3}i$
25. $-0.6840 + 1.8794i, -1.2856 - 1.5321i, 1.9696 - 0.3473i$ **27.** 2 cis 30°, 2 cis 90°,
2 cis 150°, 2 cis 210°, 2 cis 270°, 2 cis 330°   **29.** cis 0°, cis 40°, cis 80°, cis 120°,
cis 160°, cis 200°, cis 240°, cis 280°, cis 320°   **31.** cis 9°, cis 45°, cis 81°, cis 117°,
cis 153°, cis 189°, cis 225°, cis 261°, cis 297°, cis 333°   **33.** 4 cis 30°, 4 cis 270°, 4 cis 150°

PROBLEM SET 6.5 (PAGE 229)

1. $(2\sqrt{2}, 2\sqrt{2})$ **3.** $(-\frac{5}{2}, \frac{5}{2}\sqrt{3})$ **5.** $(-\frac{3}{4}\sqrt{3}, -\frac{3}{4})$ **7.** $(-2\sqrt{3}, -2)$
9. $(-2\sqrt{2}, -2\sqrt{2})$ **11.** $(5\sqrt{2}, \pi/4)$ or $(-5\sqrt{2}, 5\pi/4)$ **13.** $(4, 5\pi/3)$ or $(-4, 2\pi/3)$
15. $(3\sqrt{2}, 7\pi/4)$ or $(-3\sqrt{2}, 3\pi/4)$ **17.** $(2, 5\pi/6)$ or $(-2, 11\pi/6)$ **19.** $(13, 2.7468)$
or $(-13, 5.8985)$ **23.** Yes   **25.** No   **27.** Yes   **29.** Yes   **31.** Yes
45. $d = \sqrt{58 - 42 \cos \pi/12}$ or 4.1751

PROBLEM SET 6.6 (PAGE 238)

1. Lemniscate   **3.** Circle   **5.** Lemniscate   **7.** Circle   **9.** Cardioid
11. Limacon   **13.** Spiral   **15.** One-leaved rose (circle); $a = 2$; symmetric with
respect to the x-axis.   **17.** Three-leaved rose; $a = 6$; symmetric with respect to the
x-axis   **19.** Circle with center at the pole and radius 5/2; symmetric with respect to
the x-axis, y-axis, and origin.   **21.** Three-leaved rose; $a = 2$; symmetric with respect
to the y-axis.   **23.** $d = \sqrt{r_1^2 + r_2^2 - 2r_1 r_2 \cos(\theta_1 - \theta_2)}$
25. $r^2 = a^2 - c^2 + 2cr \cos(\theta - \alpha)$

PROBLEM SET 6.7 (PAGE 241)

1. a. $-1 - 3i$ b. $-21 - 20i$ c. $4i$ d. $-\frac{1}{2} + \frac{5}{2}i$ **3.** a. $7 - 7i = 7\sqrt{2}$ cis 315°
b. $-3i = 3$ cis 270° c. $4(\cos 7\pi/4 + i \sin 7\pi/4) = -2\sqrt{2} - 2\sqrt{2}i$
d. 2 cis 150° $= -\sqrt{3} + i$ **5.** a. 8 cis 0° b. cis 154°
7. a. $(5, 5\sqrt{3} - 2\pi) = (-5, 5\sqrt{3} - \pi)$ b. $(3, 4\pi/3) = (-3, \pi/3)$ c. $(-2, 2) = (2, 2 + \pi)$
d. $(-5, 19\pi/10) = (5, 9\pi/10)$

9.

APPENDIX B
Logarithms

.

In the text we referred to inverse functions. If $y = a^x$, then the inverse is $x = b^y$. To solve for y, we define a function called the *logarithm to the base b*.

Definition: If $x = b^y$, then $y = \log_b y$.

The laws of exponents can be written in terms of logarithms.

(1) $\log_b xy = \log_b x + \log_b y$

(2) $\log_b \dfrac{x}{y} = \log_b x - \log_b y$

(3) $\log_b x^n = n \log_b x$

In trigonometry we use logarithms to the base 10 because they are the most convenient in computations. These logarithms are called *common logarithms*, and we write $\log x$ to mean $\log_{10} x$.

We usually express a logarithm as an integer called the *characteristic* plus a decimal (between 0 and 1) called the *mantissa*. The characteristic is the exponent on 10 when the number is written in scientific notation, and the mantissa is found from a table (like Table I, for instance).

USE OF LOGARITHM TABLES

The description that follows refers to a four-place table (see Table I). For other tables there may be minor differences. Table I gives mantissas of the numbers from 1.00 to 9.99 in steps of .01. The column headed N contains the first two digits of a number, and the third digit is to the right. Decimal points are omitted from the table (that is, a decimal point belongs in front of each number on the table).

EXAMPLES:

characteristic, since $456 = 4.56 \times 10^2$ in scientific notation

1. $\log 456 = 2.$

The mantissa here is located in Table I by first finding 45 under the column headed N. Next, move over directly to the right until you are under the column headed 6: 6590.

Thus, $\log 456 = 2.6590$.

2. $\log 0.00000792 = 4.$ -10

The characteristic is -6, since $0.00000792 = 7.92 \times 10^{-6}$.

However, it is customary to express this number as $4 - 10$.

$= 4.8987 - 10$

mantissa from Table I

3. $\log 7284 = 3.$

mantissa

characteristic

Since this number has four significant digits, we would need a five-place table or else we would resort to a procedure called *interpolation*. The number 7284 is 0.4 of the way from 7280 to 7290.

Number	Mantissa	
7280	8621	← From Table I
7284		← This is what we want.
7290	8627	← From Table I

Difference is 6.

$0.4 \times 6 = 2.4$ Round this off and add to the smaller mantissa:

$8621 + 2 = 8623$.

Thus, $\log 7284 = 3.8623$.

USING LOGARITHMS FOR CALCULATIONS

We use the three properties of logarithms to simplify complicated calculations, as shown by the following example.

Find $N = \sqrt{\dfrac{(4287)(3792)}{5620}}$.

$\log N = \log\left[\dfrac{(4287)(3792)}{5620}\right]^{1/2}$

$\qquad = \dfrac{1}{2}\log\left[\dfrac{(4287)(3792)}{5620}\right]$ Property (3)

$\qquad = \dfrac{1}{2}[\log (4287)(3792) - \log 5620]$ Property (2)

$\qquad = \dfrac{1}{2}[\log 4287 + \log 3792 - \log 5620]$ Property (1)

$\log 4287 = 3.6322$

$$
\begin{array}{ccc}
 & Number & Mantissa \\
10\left[\,7\left[\begin{array}{c}4280\\4287\\4290\end{array}\right.\right. & \begin{array}{c}6314\\ \\6325\end{array} & \left.\left.]\,?\right]\,11\right.
\end{array}
$$

$0.7 \times 11 = 7.7;\ 6314 + 8 = 6322$

$+\log 3792 = 3.5788$

$\underline{}$

$\qquad\qquad 7.2110$

$$
\begin{array}{ccc}
 & Number & Mantissa \\
10\left[\,2\left[\begin{array}{c}3790\\3792\\3800\end{array}\right.\right. & \begin{array}{c}5786\\ \\5798\end{array} & \left.\left.]\,?\right]\,12\right.
\end{array}
$$

$0.2 \times 12 = 2.4;\ 5786 + 2 = 5788$

$-\log 5620 = 3.7497$

$\underline{}$

$\qquad\qquad 3.4613$

$\log N = \dfrac{1}{2}(3.4613)$

$\qquad = 1.73065$

We now find 73065 within the table of mantissas.

$$
\begin{array}{ccc}
 & Mantissa & Number \\
80\left[\,65\left[\begin{array}{c}73000\\73065\\73080\end{array}\right.\right. & \begin{array}{c}5370\\ \\5380\end{array} & \left.\left.]\,?\right]\,10\right.
\end{array}
$$

$\dfrac{?}{10} = \dfrac{65}{80}$

$? = 8$ (nearest unit)

Thus, $N = 5.378 \times 10^1$ or 53.78.

APPENDIX C
Tables

TABLE I Logarithms of Numbers.

N	0	1	2	3	4	5	6	7	8	9
10	0000	0043	0086	0128	0170	0212	0253	0294	0334	0374
11	0414	0453	0492	0531	0569	0607	0645	0682	0719	0755
12	0792	0828	0864	0899	0934	0969	1004	1038	1072	1106
13	1139	1173	1206	1239	1271	1303	1335	1367	1399	1430
14	1461	1492	1523	1553	1584	1614	1644	1673	1703	1732
15	1761	1790	1818	1847	1875	1903	1931	1959	1987	2014
16	2041	2068	2095	2122	2148	2175	2201	2227	2253	2279
17	2304	2330	2355	2380	2405	2430	2455	2480	2504	2529
18	2553	2577	2601	2625	2648	2672	2695	2718	2742	2765
19	2788	2810	2833	2856	2878	2900	2923	2945	2967	2989
20	3010	3032	3054	3075	3096	3118	3139	3160	3181	3201
21	3222	3243	3263	3284	3304	3324	3345	3365	3385	3404
22	3424	3444	3464	3483	3502	3522	3541	3560	3579	3598
23	3617	3636	3655	3674	3692	3711	3729	3747	3766	3784
24	3802	3820	3838	3856	3874	3892	3909	3927	3945	3962
25	3979	3997	4014	4031	4048	4065	4082	4099	4116	4133
26	4150	4166	4183	4200	4216	4232	4249	4265	4281	4298
27	4314	4330	4346	4362	4378	4393	4409	4425	4440	4456
28	4472	4487	4502	4518	4533	4548	4564	4579	4594	4609
29	4624	4639	4654	4669	4683	4698	4713	4728	4742	4757
30	4771	4786	4800	4814	4829	4843	4857	4871	4886	4900
31	4914	4928	4942	4955	4969	4983	4997	5011	5024	5038
32	5051	5065	5079	5092	5105	5119	5132	5145	5159	5172
33	5185	5198	5211	5224	5237	5250	5263	5276	5289	5302
34	5315	5328	5340	5353	5366	5378	5391	5403	5416	5428
35	5441	5453	5465	5478	5490	5502	5514	5527	5539	5551
36	5563	5575	5587	5599	5611	5623	5635	5647	5658	5670
37	5682	5694	5705	5717	5729	5740	5752	5763	5775	5786
38	5798	5809	5821	5832	5843	5855	5866	5877	5888	5899
39	5911	5922	5933	5944	5955	5966	5977	5988	5999	6010
40	6021	6031	6042	6053	6064	6075	6085	6096	6107	6117
41	6128	6138	6149	6160	6170	6180	6191	6201	6212	6222
42	6232	6243	6253	6263	6274	6284	6294	6304	6314	6325
43	6335	6345	6355	6365	6375	6385	6395	6405	6415	6425
44	6435	6444	6454	6464	6474	6484	6493	6503	6513	6522
45	6532	6542	6551	6561	6571	6580	6590	6599	6609	6618
46	6628	6637	6646	6656	6665	6675	6684	6693	6702	6712
47	6721	6730	6739	6749	6758	6767	6776	6785	6794	6803
48	6812	6821	6830	6839	6848	6857	6866	6875	6884	6893
49	6902	6911	6920	6928	6937	6946	6955	6964	6972	6981
50	6990	6998	7007	7016	7024	7033	7042	7050	7059	7067
51	7076	7084	7093	7101	7110	7118	7126	7135	7143	7152
52	7160	7168	7177	7185	7193	7202	7210	7218	7226	7235
53	7243	7251	7259	7267	7275	7284	7292	7300	7308	7316
54	7324	7332	7340	7348	7356	7364	7372	7380	7388	7396
N	0	1	2	3	4	5	6	7	8	9

TABLE I   Logarithms of Numbers (*continued*).

N	0	1	2	3	4	5	6	7	8	9
55	7404	7412	7419	7427	7435	7443	7451	7459	7466	7474
56	7482	7490	7497	7505	7513	7520	7528	7536	7543	7551
57	7559	7566	7574	7582	7589	7597	7604	7612	7619	7627
58	7634	7642	7649	7657	7664	7672	7679	7686	7694	7701
59	7709	7716	7723	7731	7738	7745	7752	7760	7767	7774
60	7782	7789	7796	7803	7810	7818	7825	7832	7839	7846
61	7853	7860	7868	7875	7882	7889	7896	7903	7910	7917
62	7924	7931	7938	7945	7952	7959	7966	7973	7980	7987
63	7993	8000	8007	8014	8021	8028	8035	8041	8048	8055
64	8062	8069	8075	8082	8089	8096	8102	8109	8116	8122
65	8129	8136	8142	8149	8156	8162	8169	8176	8182	8189
66	8195	8202	8209	8215	8222	8228	8235	8241	8248	8254
67	8261	8267	8274	8280	8287	8293	8299	8306	8312	8319
68	8325	8331	8338	8344	8351	8357	8363	8370	8376	8382
69	8388	8395	8401	8407	8414	8420	8426	8432	8439	8445
70	8451	8457	8463	8470	8476	8482	8488	8494	8500	8506
71	8513	8519	8525	8531	8537	8543	8549	8555	8561	8567
72	8573	8579	8585	8591	8597	8603	8609	8615	8621	8627
73	8633	8639	8645	8651	8657	8663	8669	8675	8681	8686
74	8692	8698	8704	8710	8716	8722	8727	8733	8739	8745
75	8751	8756	8762	8768	8774	8779	8785	8791	8797	8802
76	8808	8814	8820	8825	8831	8837	8842	8848	8854	8859
77	8865	8871	8876	8882	8887	8893	8899	8904	8910	8915
78	8921	8927	8932	8938	8943	8949	8954	8960	8965	8971
79	8976	8982	8987	8993	8998	9004	9009	9015	9020	9025
80	9031	9036	9042	9047	9053	9058	9063	9069	9074	9079
81	9085	9090	9096	9101	9106	9112	9117	9122	9128	9133
82	9138	9143	9149	9154	9159	9165	9170	9175	9180	9186
83	9191	9196	9201	9206	9212	9217	9222	9227	9232	9238
84	9243	9248	9253	9258	9263	9269	9274	9279	9284	9289
85	9294	9299	9304	9309	9315	9320	9325	9330	9335	9340
86	9345	9350	9355	9360	9365	9370	9375	9380	9385	9390
87	9395	9400	9405	9410	9415	9420	9425	9430	9435	9440
88	9445	9450	9455	9460	9465	9469	9474	9479	9484	9489
89	9494	9499	9504	9509	9513	9518	9523	9528	9533	9538
90	9542	9547	9552	9557	9562	9566	9571	9576	9581	9586
91	9590	9595	9600	9605	9609	9614	9619	9624	9628	9633
92	9638	9643	9647	9652	9657	9661	9666	9671	9675	9680
93	9685	9689	9694	9699	9703	9708	9713	9717	9722	9727
94	9731	9736	9741	9745	9750	9754	9759	9763	9768	9773
95	9777	9782	9786	9791	9795	9800	9805	9809	9814	9818
96	9823	9827	9832	9836	9841	9845	9850	9854	9859	9863
97	9868	9872	9877	9881	9886	9890	9894	9899	9903	9908
98	9912	9917	9921	9926	9930	9934	9939	9943	9948	9952
99	9956	9961	9965	9969	9974	9978	9983	9987	9991	9996
N	0	1	2	3	4	5	6	7	8	9

TABLE II   Powers and Roots.

n	n^2	$\sqrt{n}$	$\sqrt{10n}$	n^3	$\sqrt[3]{n}$	$\sqrt[3]{10n}$	$\sqrt[3]{100n}$
1	1	1.000 000	3.162 278	1	1.000 000	2.154 435	4.641 589
2	4	1.414 214	4.472 136	8	1.259 921	2.714 418	5.848 035
3	9	1.732 051	5.477 226	27	1.442 250	3.107 233	6.694 330
4	16	2.000 000	6.324 555	64	1.587 401	3.419 952	7.368 063
5	25	2.236 068	7.071 068	125	1.709 976	3.684 031	7.937 005
6	36	2.449 490	7.745 967	216	1.817 121	3.914 868	8.434 327
7	49	2.645 751	8.366 600	343	1.912 931	4.121 285	8.879 040
8	64	2.828 427	8.944 272	512	2.000 000	4.308 869	9.283 178
9	81	3.000 000	9.486 833	729	2.080 084	4.481 405	9.654 894
10	100	3.162 278	10.00000	1 000	2.154 435	4.641 589	10.00000
11	121	3.316 625	10.48809	1 331	2.223 980	4.791 420	10.32280
12	144	3.464 102	10.95445	1 728	2.289 428	4.932 424	10.62659
13	169	3.605 551	11.40175	2 197	2.351 335	5.065 797	10.91393
14	196	3.741 657	11.83216	2 744	2.410 142	5.192 494	11.18689
15	225	3.872 983	12.24745	3 375	2.466 212	5.313 293	11.44714
16	256	4.000 000	12.64911	4 096	2.519 842	5.428 835	11.69607
17	289	4.123 106	13.03840	4 913	2.571 282	5.539 658	11.93483
18	324	4.242 641	13.41641	5 832	2.620 741	5.646 216	12.16440
19	361	4.358 899	13.78405	6 859	2.668 402	5.748 897	12.38562
20	400	4.472 136	14.14214	8 000	2.714 418	5.848 035	12.59921
21	441	4.582 576	14.49138	9 261	2.758 924	5.943 922	12.80579
22	484	4.690 416	14.83240	10 648	2.802 039	6.036 811	13.00591
23	529	4.795 832	15.16575	12 167	2.843 867	6.126 926	13.20006
24	576	4.898 979	15.49193	13 824	2.884 499	6.214 465	13.38866
25	625	5.000 000	15.81139	15 625	2.924 018	6.299 605	13.57209
26	676	5.099 020	16.12452	17 576	2.962 496	6.382 504	13.75069
27	729	5.196 152	16.43168	19 683	3.000 000	6.463 304	13.92477
28	784	5.291 503	16.73320	21 952	3.036 589	6.542 133	14.09460
29	841	5.385 165	17.02939	24 389	3.072 317	6.619 106	14.26043
30	900	5.477 226	17.32051	27 000	3.107 233	6.694 330	14.42250
31	961	5.567 764	17.60682	29 791	3.141 381	6.767 899	14.58100
32	1 024	5.656 854	17.88854	32 768	3.174 802	6.839 904	14.73613
33	1 089	5.744 563	18.16590	35 937	3.207 534	6.910 423	14.88806
34	1 156	5.830 952	18.43909	39 304	3.239 612	6.979 532	15.03695
35	1 225	5.916 080	18.70829	42 875	3.271 066	7.047 299	15.18294
36	1 296	6.000 000	18.97367	46 656	3.301 927	7.113 787	15.32619
37	1 369	6.082 763	19.23538	50 653	3.332 222	7.179 054	15.46680
38	1 444	6.164 414	19.49359	54 872	3.361 975	7.243 156	15.60491
39	1 521	6.244 998	19.74842	59 319	3.391 211	7.306 144	15.74061
40	1 600	6.324 555	20.00000	64 000	3.419 952	7.368 063	15.87401
41	1 681	6.403 124	20.24846	68 921	3.448 217	7.428 959	16.00521
42	1 764	6.480 741	20.49390	74 088	3.476 027	7.488 872	16.13429
43	1 849	6.557 439	20.73644	79 507	3.503 398	7.547 842	16.26133
44	1 936	6.633 250	20.97618	85 184	3.530 348	7.605 905	16.38643
45	2 025	6.708 204	21.21320	91 125	3.556 893	7.663 094	16.50964
46	2 116	6.782 330	21.44761	97 336	3.583 048	7.719 443	16.63103
47	2 209	6.855 655	21.67948	103 823	3.608 826	7.774 980	16.75069
48	2 304	6.928 203	21.90890	110 592	3.634 241	7.829 735	16.86865
49	2 401	7.000 000	22.13594	117 649	3.659 306	7.883 735	16.98499
50	2 500	7.071 068	22.36068	125 000	3.684 031	7.937 005	17.09976

TABLE II Powers and Roots (*continued*).

n	n^2	$\sqrt{n}$	$\sqrt{10n}$	n^3	$\sqrt[3]{n}$	$\sqrt[3]{10n}$	$\sqrt[3]{100n}$
50	2 500	7.071 068	22.36068	125 000	3.684 031	7.937 005	17.09976
51	2 601	7.141 428	22.58318	132 651	3.708 430	7.989 570	17.21301
52	2 704	7.211 103	22.80351	140 608	3.732 511	8.041 452	17.32478
53	2 809	7.280 110	23.02173	148 877	3.756 286	8.092 672	17.43513
54	2 916	7.348 469	23.23790	157 464	3.779 763	8.143 253	17.54411
55	3 025	7.416 198	23.45208	166 375	3.802 952	8.193 213	17.65174
56	3 136	7.483 315	23.66432	175 616	3.825 862	8.242 571	17.75808
57	3 249	7.549 834	23.87467	185 193	3.848 501	8.291 344	17.86316
58	3 364	7.615 773	24.08319	195 112	3.870 877	8.339 551	17.96702
59	3 481	7.681 146	24.28992	205 379	3.892 996	8.387 207	18.06969
60	3 600	7.745 967	24.49490	216 000	3.914 868	8.434 327	18.17121
61	3 721	7.810 250	24.69818	226 981	3.936 497	8.480 926	18.27160
62	3 844	7.874 008	24.89980	238 328	3.957 892	8.527 019	18.37091
63	3 969	7.937 254	25.09980	250 047	3.979 057	8.572 619	18.46915
64	4 096	8.000 000	25.29822	262 144	4.000 000	8.617 739	18.56636
65	4 225	8.062 258	25.49510	274 625	4.020 726	8.662 391	18.66256
66	4 356	8.124 038	25.69047	287 496	4.041 240	8.706 588	18.75777
67	4 489	8.185 353	25.88436	300 763	4.061 548	8.750 340	18.85204
68	4 624	8.246 211	26.07681	314 432	4.081 655	8.793 659	18.94536
69	4 761	8.306 624	26.26785	328 509	4.101 566	8.836 556	19.03778
70	4 900	8.366 600	26.45751	343 000	4.121 285	8.879 040	19.12931
71	5 041	8.426 150	26.64583	357 911	4.140 818	8.921 121	19.21997
72	5 184	8.485 281	26.83282	373 248	4.160 168	8.962 809	19.30979
73	5 329	8.544 004	27.01851	389 017	4.179 339	9.004 113	19.39877
74	5 476	8.602 325	27.20294	405 224	4.198 336	9.045 042	19.48695
75	5 625	8.660 254	27.38613	421 875	4.217 163	9.085 603	19.57434
76	5 776	8.717 798	27.56810	438 976	4.235 824	9.125 805	19.66095
77	5 929	8.774 964	27.74887	456 533	4.254 321	9.165 656	19.74681
78	6 084	8.831 761	27.92848	474 552	4.272 659	9.205 164	19.83192
79	6 241	8.888 194	28.10694	493 039	4.290 840	9.244 335	19.91632
80	6 400	8.944 272	28.28427	512 000	4.308 869	9.283 178	20.00000
81	6 561	9.000 000	28.46050	531 441	4.326 749	9.321 698	20.08299
82	6 724	9.055 385	28.63564	551 368	4.344 481	9.359 902	20.16530
83	6 889	9.110 434	28.80972	571 787	4.362 071	9.397 796	20.24694
84	7 056	9.165 151	28.98275	592 704	4.379 519	9.435 388	20.32793
85	7 225	9.219 544	29.15476	614 125	4.396 830	9.472 682	20.40828
86	7 396	9.273 618	29.32576	636 056	4.414 005	9.509 685	20.48800
87	7 569	9.327 379	29.49576	658 503	4.431 048	9.546 403	20.56710
88	7 744	9.380 832	29.66479	681 472	4.447 960	9.582 840	20.64560
89	7 921	9.433 981	29.83287	704 969	4.464 745	9.619 002	20.72351
90	8 100	9.486 833	30.00000	729 000	4.481 405	9.654 894	20.80084
91	8 281	9.539 392	30.16621	753 571	4.497 941	9.690 521	20.87759
92	8 464	9.591 663	30.33150	778 688	4.514 357	9.725 888	20.95379
93	8 649	9.643 651	30.49590	804 357	4.530 655	9.761 000	21.02944
94	8 836	9.695 360	30.65942	830 584	4.546 836	9.795 861	21.10454
95	9 025	9.746 794	30.82207	857 375	4.562 903	9.830 476	21.17912
96	9 216	9.797 959	30.98387	884 736	4.578 857	9.864 848	21.25317
97	9 409	9.848 858	31.14482	912 673	4.594 701	9.898 983	21.32671
98	9 604	9.899 495	31.30495	941 192	4.610 436	9.932 884	21.39975
99	9 801	9.949 874	31.46427	970 299	4.626 065	9.966 555	21.47229
100	10 000	10.00000	31.62278	1 000 000	4.641 589	10.00000	21.54435

TABLE III   Trigonometric (Degrees).

Deg.	Sin	Tan	*Cot	Cos	
0.0	0.00000	0.00000	∞	1.0000	**90.0**
.1	.00175	.00175	573.0	1.0000	89.9
.2	.00349	.00349	286.5	1.0000	.8
.3	.00524	.00524	191.0	1.0000	.7
.4	.00698	.00698	143.24	1.0000	.6
.5	.00873	.00873	114.59	1.0000	.5
.6	.01047	.01047	95.49	0.9999	.4
.7	.01222	.01222	81.85	.9999	.3
.8	.01396	.01396	71.62	.9999	.2
.9	.01571	.01571	63.66	.9999	89.1
1.0	0.01745	0.01746	57.29	0.9998	**89.0**
.1	.01920	.01920	52.08	.9998	88.9
.2	.02094	.02095	47.74	.9998	.8
.3	.02269	.02269	44.07	.9997	.7
.4	.02443	.02444	40.92	.9997	.6
.5	.02618	.02619	38.19	.9997	.5
.6	.02792	.02793	35.80	.9996	.4
.7	.02967	.02968	33.69	.9996	.3
.8	.03141	.03143	31.82	.9995	.2
.9	.03316	.03317	30.14	.9995	88.1
2.0	0.03490	0.03492	28.64	0.9994	**88.0**
.1	.03664	.03667	27.27	.9993	87.9
.2	.03839	.03842	26.03	.9993	.8
.3	.04013	.04016	24.90	.9992	.7
.4	.04188	.04191	23.86	.9991	.6
.5	.04362	.04366	22.90	.9990	.5
.6	.04536	.04541	22.02	.9990	.4
.7	.04711	.04716	21.20	.9989	.3
.8	.04885	.04891	20.45	.9988	.2
.9	.05059	.05066	19.74	.9987	87.1
3.0	0.05234	0.05241	19.081	0.9986	**87.0**
.1	.05408	.05416	18.464	.9985	86.9
.2	.05582	.05591	17.886	.9984	.8
.3	.05756	.05766	17.343	.9983	.7
.4	.05931	.05941	16.832	.9982	.6
.5	.06105	.06116	16.350	.9981	.5
.6	.06279	.06291	15.895	.9980	.4
.7	.06453	.06467	15.464	.9979	.3
.8	.06627	.06642	15.056	.9978	.2
.9	.06802	.06817	14.669	.9977	86.1
4.0	0.06976	0.06993	14.301	0.9976	**86.0**
.1	.07150	.07168	13.951	.9974	85.9
.2	.07324	.07344	13.617	.9973	.8
.3	.07498	.07519	13.300	.9972	.7
.4	.07672	.07695	12.996	.9971	.6
.5	.07846	.07870	12.706	.9969	.5
.6	.08020	.08046	12.429	.9968	.4
.7	.08194	.08221	12.163	.9966	.3
.8	.08368	.08397	11.909	.9965	.2
.9	.08542	.08573	11.664	.9963	85.1
5.0	0.08716	0.08749	11.430	0.9962	**85.0**
.1	.08889	.08925	11.205	.9960	84.9
.2	.09063	.09101	10.988	.9959	.8
.3	.09237	.09277	10.780	.9957	.7
.4	.09411	.09453	10.579	.9956	.6
.5	.09585	.09629	10.385	.9954	.5
.6	.09758	.09805	10.199	.9952	.4
.7	.09932	.09981	10.019	.9951	.3
.8	.10106	.10158	9.845	.9949	.2
.9	.10279	.10334	9.677	.9947	84.1
6.0	0.10453	0.10510	9.514	0.9945	**84.0**
	Cos	Cot	*Tan	Sin	Deg.

Deg.	Sin	Tan	Cot	Cos	
6.0	0.10453	0.10510	9.514	0.9945	**84.0**
.1	.10626	.10687	9.357	.9943	83.9
.2	.10800	.10863	9.205	.9942	.8
.3	.10973	.11040	9.058	.9940	.7
.4	.11147	.11217	8.915	.9938	.6
.5	.11320	.11394	8.777	.9936	.5
.6	.11494	.11570	8.643	.9934	.4
.7	.11667	.11747	8.513	.9932	.3
.8	.11840	.11924	8.386	.9930	.2
.9	.12014	.12101	8.264	.9928	83.1
7.0	0.12187	0.12278	8.144	0.9925	**83.0**
.1	.12360	.12456	8.028	.9923	82.9
.2	.12533	.12633	7.916	.9921	.8
.3	.12706	.12810	7.806	.9919	.7
.4	.12880	.12988	7.700	.9917	.6
.5	.13053	.13165	7.596	.9914	.5
.6	.13226	.13343	7.495	.9912	.4
.7	.13399	.13521	7.396	.9910	.3
.8	.13572	.13698	7.300	.9907	.2
.9	.13744	.13876	7.207	.9905	82.1
8.0	0.13917	0.14054	7.115	0.9903	**82.0**
.1	.14090	.14232	7.026	.9900	81.9
.2	.14263	.14410	6.940	.9898	.8
.3	.14436	.14588	6.855	.9895	.7
.4	.14608	.14767	6.772	.9893	.6
.5	.14781	.14945	6.691	.9890	.5
.6	.14954	.15124	6.612	.9888	.4
.7	.15126	.15302	6.535	.9885	.3
.8	.15299	.15481	6.460	.9882	.2
.9	.15471	.15660	6.386	.9880	81.1
9.0	0.15643	0.15838	6.314	0.9877	**81.0**
.1	.15816	.16017	6.243	.9874	80.9
.2	.15988	.16196	6.174	.9871	.8
.3	.16160	.16376	6.107	.9869	.7
.4	.16333	.16555	6.041	.9866	.6
.5	.16505	.16734	5.976	.9863	.5
.6	.16677	.16914	5.912	.9860	.4
.7	.16849	.17093	5.850	.9857	.3
.8	.17021	.17273	5.789	.9854	.2
.9	.17193	.17453	5.730	.9851	80.1
10.0	0.1736	0.1763	5.671	0.9848	**80.0**
.1	.1754	.1781	5.614	.9845	79.9
.2	.1771	.1799	5.558	.9842	.8
.3	.1788	.1817	5.503	.9839	.7
.4	.1805	.1835	5.449	.9836	.6
.5	.1822	.1853	5.396	.9833	.5
.6	.1840	.1871	5.343	.9829	.4
.7	.1857	.1890	5.292	.9826	.3
.8	.1874	.1908	5.242	.9823	.2
.9	.1891	.1926	5.193	.9820	79.1
11.0	0.1908	0.1944	5.145	0.9816	**79.0**
.1	.1925	.1962	5.097	.9813	78.9
.2	.1942	.1980	5.050	.9810	.8
.3	.1959	.1998	5.005	.9806	.7
.4	.1977	.2016	4.959	.9803	.6
.5	.1994	.2035	4.915	.9799	.5
.6	.2011	.2053	4.872	.9796	.4
.7	.2028	.2071	4.829	.9792	.3
.8	.2045	.2089	4.787	.9789	.2
.9	.2062	.2107	4.745	.9785	78.1
12.0	0.2079	0.2126	4.705	0.9781	**78.0**
	Cos	Cot	Tan	Sin	Deg.

*Interpolation in this section of the table is inaccurate.

TABLE III   Trigonometric (Degrees) (*continued*).

Deg.	Sin	Tan	Cot	Cos	
12.0	0.2079	0.2126	4.705	0.9781	**78.0**
.1	.2096	.2144	4.665	.9778	77.9
.2	.2113	.2162	4.625	.9774	.8
.3	.2130	.2180	4.586	.9770	.7
.4	.2147	.2199	4.548	.9767	.6
.5	.2164	.2217	4.511	.9763	.5
.6	.2181	.2235	4.474	.9759	.4
.7	.2198	.2254	4.437	.9755	.3
.8	.2215	.2272	4.402	.9751	.2
.9	.2233	.2290	4.366	.9748	77.1
13.0	0.2250	0.2309	4.331	0.9744	**77.0**
.1	.2267	.2327	4.297	.9740	76.9
.2	.2284	.2345	4.264	.9736	.8
.3	.2300	.2364	4.230	.9732	.7
.4	.2317	.2382	4.198	.9728	.6
.5	.2334	.2401	4.165	.9724	.5
.6	.2351	.2419	4.134	.9720	.4
.7	.2368	.2438	4.102	.9715	.3
.8	.2385	.2456	4.071	.9711	.2
.9	.2402	.2475	4.041	.9707	76.1
14.0	0.2419	0.2493	4.011	0.9703	**76.0**
.1	.2436	.2512	3.981	.9699	75.9
.2	.2453	.2530	3.952	.9694	.8
.3	.2470	.2549	3.923	.9690	.7
.4	.2487	.2568	3.895	.9686	.6
.5	.2504	.2586	3.867	.9681	.5
.6	.2521	.2605	3.839	.9677	.4
.7	.2538	.2623	3.812	.9673	.3
.8	.2554	.2642	3.785	.9668	.2
.9	.2571	.2661	3.758	.9664	75.1
15.0	0.2588	0.2679	3.732	0.9659	**75.0**
.1	.2605	.2698	3.706	.9655	74.9
.2	.2622	.2717	3.681	.9650	.8
.3	.2639	.2736	3.655	.9646	.7
.4	.2656	.2754	3.630	.9641	.6
.5	.2672	.2773	3.606	.9636	.5
.6	.2689	.2792	3.582	.9632	.4
.7	.2706	.2811	3.558	.9627	.3
.8	.2723	.2830	3.534	.9622	.2
.9	.2740	.2849	3.511	.9617	74.1
16.0	0.2756	0.2867	3.487	0.9613	**74.0**
.1	.2773	.2886	3.465	.9608	73.9
.2	.2790	.2905	3.442	.9603	.8
.3	.2807	.2924	3.420	.9598	.7
.4	.2823	.2943	3.398	.9593	.6
.5	.2840	.2962	3.376	.9588	.5
.6	.2857	.2981	3.354	.9583	.4
.7	.2874	.3000	3.333	.9578	.3
.8	.2890	.3019	3.312	.9573	.2
.9	.2907	.3038	3.291	.9568	73.1
17.0	0.2924	0.3057	3.271	0.9563	**73.0**
.1	.2940	.3076	3.251	.9558	72.9
.2	.2957	.3096	3.230	.9553	.8
.3	.2974	.3115	3.211	.9548	.7
.4	.2990	.3134	3.191	.9542	.6
.5	.3007	.3153	3.172	.9537	.5
.6	.3024	.3172	3.152	.9532	.4
.7	.3040	.3191	3.133	.9527	.3
.8	.3057	.3211	3.115	.9521	.2
.9	.3074	.3230	3.096	.9516	72.1
18.0	0.3090	0.3249	3.078	0.9511	**72.0**
	Cos	Cot	Tan	Sin	Deg.

Deg.	Sin	Tan	Cot	Cos	
18.0	0.3090	0.3249	3.078	0.9511	**72.0**
.1	.3107	.3269	3.060	.9505	71.9
.2	.3123	.3288	3.042	.9500	.8
.3	.3140	.3307	3.024	.9494	.7
.4	.3156	.3327	3.006	.9489	.6
.5	.3173	.3346	2.989	.9483	.5
.6	.3190	.3365	2.971	.9478	.4
.7	.3206	.3385	2.954	.9472	.3
.8	.3223	.3404	2.937	.9466	.2
.9	.3239	.3424	2.921	.9461	71.1
19.0	0.3256	0.3443	2.904	0.9455	**71.0**
.1	.3272	.3463	2.888	.9449	70.9
.2	.3289	.3482	2.872	.9444	.8
.3	.3305	.3502	2.856	.9438	.7
.4	.3322	.3522	2.840	.9432	.6
.5	.3338	.3541	2.824	.9426	.5
.6	.3355	.3561	2.808	.9421	.4
.7	.3371	.3581	2.793	.9415	.3
.8	.3387	.3600	2.778	.9409	.2
.9	.3404	.3620	2.762	.9403	70.1
20.0	0.3420	0.3640	2.747	0.9397	**70.0**
.1	.3437	.3659	2.733	.9391	69.9
.2	.3453	.3679	2.718	.9385	.8
.3	.3469	.3699	2.703	.9379	.7
.4	.3486	.3719	2.689	.9373	.6
.5	.3502	.3739	2.675	.9367	.5
.6	.3518	.3759	2.660	.9361	.4
.7	.3535	.3779	2.646	.9354	.3
.8	.3551	.3799	2.633	.9348	.2
.9	.3567	.3819	2.619	.9342	69.1
21.0	0.3584	0.3839	2.605	0.9336	**69.0**
.1	.3600	.3859	2.592	.9330	68.9
.2	.3616	.3879	2.578	.9323	.8
.3	.3633	.3899	2.565	.9317	.7
.4	.3649	.3919	2.552	.9311	.6
.5	.3665	.3939	2.539	.9304	.5
.6	.3681	.3959	2.526	.9298	.4
.7	.3697	.3979	2.513	.9291	.3
.8	.3714	.4000	2.500	.9285	.2
.9	.3730	.4020	2.488	.9278	68.1
22.0	0.3746	0.4040	2.475	0.9272	**68.0**
.1	.3762	.4061	2.463	.9265	67.9
.2	.3778	.4081	2.450	.9259	.8
.3	.3795	.4101	2.438	.9252	.7
.4	.3811	.4122	2.426	.9245	.6
.5	.3827	.4142	2.414	.9239	.5
.6	.3843	.4163	2.402	.9232	.4
.7	.3859	.4183	2.391	.9225	.3
.8	.3875	.4204	2.379	.9219	.2
.9	.3891	.4224	2.367	.9212	67.1
23.0	0.3907	0.4245	2.356	0.9205	**67.0**
.1	.3923	.4265	2.344	.9198	66.9
.2	.3939	.4286	2.333	.9191	.8
.3	.3955	.4307	2.322	.9184	.7
.4	.3971	.4327	2.311	.9178	.6
.5	.3987	.4348	2.300	.9171	.5
.6	.4003	.4369	2.289	.9164	.4
.7	.4019	.4390	2.278	.9157	.3
.8	.4035	.4411	2.267	.9150	.2
.9	.4051	.4431	2.257	.9143	66.1
24.0	0.4067	0.4452	2.246	0.9135	**66.0**
	Cos	Cot	Tan	Sin	Deg.

TABLE III   Trigonometric (Degrees) (*continued*).

Deg.	Sin	Tan	Cot	Cos	
24.0	0.4067	0.4452	2.246	0.9135	**66.0**
.1	.4083	.4473	2.236	.9128	65.9
.2	.4099	.4494	2.225	.9121	.8
.3	.4115	.4515	2.215	.9114	.7
.4	.4131	.4536	2.204	.9107	.6
.5	.4147	.4557	2.194	.9100	.5
.6	.4163	.4578	2.184	.9092	.4
.7	.4179	.4599	2.174	.9085	.3
.8	.4195	.4621	2.164	.9078	.2
.9	.4210	.4642	2.154	.9070	65.1
25.0	0.4226	0.4663	2.145	0.9063	**65.0**
.1	.4242	.4684	2.135	.9056	64.9
.2	.4258	.4706	2.125	.9048	.8
.3	.4274	.4727	2.116	.9041	.7
.4	.4289	.4748	2.106	.9033	.6
.5	.4305	.4770	2.097	.9026	.5
.6	.4321	.4791	2.087	.9018	.4
.7	.4337	.4813	2.078	.9011	.3
.8	.4352	.4834	2.069	.9003	.2
.9	.4368	.4856	2.059	.8996	64.1
26.0	0.4384	0.4877	2.050	0.8988	**64.0**
.1	.4399	.4899	2.041	.8980	63.9
.2	.4415	.4921	2.032	.8973	.8
.3	.4431	.4942	2.023	.8965	.7
.4	.4446	.4964	2.014	.8957	.6
.5	.4462	.4986	2.006	.8949	.5
.6	.4478	.5008	1.997	.8942	.4
.7	.4493	5029	1.988	.8934	.3
.8	.4509	.5051	1.980	.8926	.2
.9	.4524	.5073	1.971	.8918	63.1
27.0	0.4540	0.5095	1.963	0.8910	**63.0**
.1	.4555	.5117	1.954	.8902	62.9
.2	.4571	.5139	1.946	.8894	.8
.3	.4586	5161	1.937	.8886	.7
.4	.4602	.5184	1.929	.8878	.6
.5	.4617	.5206	1 021	.8870	.5
.6	.4633	.5228	1.913	.8862	.4
.7	.4648	.5250	1.905	.8854	.3
.8	.4664	.5272	1.897	.8846	.2
.9	.4679	.5295	1.889	.8838	62.1
28.0	0.4695	0.5317	1.881	0.8829	**62.0**
.1	.4710	.5340	1.873	.8821	61.9
.2	.4726	.5362	1.865	.8813	.8
.3	.4741	.5384	1.857	.8805	.7
.4	.4756	.5407	1.849	.8796	.6
.5	.4772	.5430	1.842	.8788	.5
.6	.4787	.5452	1.834	.8780	.4
.7	.4802	.5475	1.827	.8771	.3
.8	.4818	.5498	1.819	.8763	.2
.9	.4833	.5520	1.811	.8755	61.1
29.0	0.4848	0.5543	1.804	0.8746	**61.0**
.1	.4863	.5566	1.797	.8738	60.9
.2	.4879	.5589	1.789	.8729	.8
.3	.4894	.5612	1.782	.8721	.7
.4	.4909	.5635	1.775	.8712	.6
.5	.4924	.5658	1.767	.8704	.5
.6	.4939	.5681	1.760	.8695	.4
.7	.4955	.5704	1.753	.8686	.3
.8	.4970	.5727	1.746	.8678	.2
.9	.4985	.5750	1.739	.8669	60.1
30.0	0.5000	0.5774	1.732	0.8660	**60.0**
	Cos	Cot	Tan	Sin	Deg.

Deg.	Sin	Tan	Cot	Cos	
30.0	0.5000	0.5774	1.7321	0.8660	**60.0**
.1	.5015	.5797	1.7251	.8652	59.9
.2	.5030	.5820	1.7182	.8643	.8
.3	.5045	.5844	1.7113	.8634	.7
.4	.5060	.5867	1.7045	.8625	.6
.5	.5075	.5890	1.6977	.8616	.5
.6	.5090	.5914	1.6909	.8607	.4
.7	.5105	.5938	1.6842	.8599	.3
.8	.5120	.5961	1.6775	.8590	.2
.9	.5135	.5985	1.6709	.8581	59.1
31.0	0.5150	0.6009	1.6643	0.8572	**59.0**
.1	.5165	.6032	1.6577	.8563	58.9
.2	.5180	.6056	1.6512	.8554	.8
.3	.5195	.6080	1.6447	.8545	.7
.4	.5210	.6104	1.6383	.8536	.6
.5	.5225	.6128	1.6319	.8526	.5
.6	.5240	.6152	1.6255	.8517	.4
.7	.5255	.6176	1.6191	.8508	.3
.8	.5270	.6200	1.6128	.8499	.2
.9	.5284	.6224	1.6066	.8490	58.1
32.0	0.5299	0.6249	1.6003	0.8480	**58.0**
.1	.5314	.6273	1.5941	.8471	57.9
.2	.5329	.6297	1.5880	.8462	.8
.3	.5344	.6322	1.5818	.8453	.7
.4	.5358	.6346	1.5757	.8443	.6
.5	.5373	.6371	1.5697	.8434	.5
.6	.5388	.6395	1.5637	.8425	.4
.7	.5402	.6420	1.5577	.8415	.3
.8	.5417	.6445	1.5517	.8406	.2
.9	.5432	.6469	1.5458	.8396	57.1
33.0	0.5446	0.6494	1.5399	0.8387	**57.0**
.1	.5461	.6519	1.5340	.8377	56.9
.2	.5476	.6544	1.5282	.8368	.8
.3	.5490	.6569	1.5224	.8358	.7
.4	.5505	.6594	1.5166	.8348	.6
.5	.5519	.6619	1.5108	.8339	.5
.6	.5534	.6644	1.5051	.8329	.4
.7	.5548	.6669	1.4994	.8320	.3
.8	.5563	.6694	1.4938	.8310	.2
.9	.5577	.6720	1.4882	.8300	56.1
34.0	0.5592	0.6745	1.4826	0.8290	**56.0**
.1	.5606	.6771	1.4770	.8281	55.9
.2	.5621	.6796	1.4715	.8271	.8
.3	.5635	.6822	1.4659	.8261	.7
.4	.5650	.6847	1.4605	.8251	.6
.5	.5664	.6873	1.4550	.8241	.5
.6	.5678	.6899	1.4496	.8231	.4
.7	.5693	.6924	1.4442	.8221	.3
.8	.5707	.6950	1.4388	.8211	.2
.9	.5721	.6976	1.4335	.8202	55.1
35.0	0.5736	0.7002	1.4281	0.8192	**55.0**
.1	.5750	.7028	1.4229	.8181	54.9
.2	.5764	.7054	1.4176	.8171	.8
.3	.5779	.7080	1.4124	.8161	.7
.4	.5793	.7107	1.4071	.8151	.6
.5	.5807	.7133	1.4019	.8141	.5
.6	.5821	.7159	1.3968	.8131	.4
.7	.5835	.7186	1.3916	.8121	.3
.8	.5850	.7212	1.3865	.8111	.2
.9	.5864	.7239	1.3814	.8100	54.1
36.0	0.5878	0.7265	1.3764	0.8090	**54.0**
	Cos	Cot	Tan	Sin	Deg.

TABLE III   Trigonometric (Degrees) (*continued*).

Deg.	Sin	Tan	Cot	Cos	
36.0	0.5878	0.7265	1.3764	0.8090	**54.0**
.1	.5892	.7292	1.3713	.8080	53.9
.2	.5906	.7319	1.3663	.8070	.8
.3	.5920	.7346	1.3613	.8059	.7
.4	.5934	.7373	1.3564	.8049	.6
.5	.5948	.7400	1.3514	.8039	.5
.6	.5962	.7427	1.3465	.8028	.4
.7	.5976	.7454	1.3416	.8018	.3
.8	.5990	.7481	1.3367	.8007	.2
.9	.6004	.7508	1.3319	.7997	53.1
37.0	0.6018	0.7536	1.3270	0.7986	**53.0**
.1	.6032	.7563	1.3222	.7976	52.9
.2	.6046	.7590	1.3175	.7965	.8
.3	.6060	.7618	1.3127	.7955	.7
.4	.6074	.7646	1.3079	.7944	.6
.5	.6088	.7673	1.3032	.7934	.5
.6	.6101	.7701	1.2985	.7923	.4
.7	.6115	.7729	1.2938	.7912	.3
.8	.6129	.7757	1.2892	.7902	.2
.9	.6143	.7785	1.2846	.7891	52.1
38.0	0.6157	0.7813	1.2799	0.7880	**52.0**
.1	.6170	.7841	1.2753	.7869	51.9
.2	.6184	.7869	1.2708	.7859	.8
.3	.6198	.7898	1.2662	.7848	.7
.4	.6211	.7926	1.2617	.7837	.6
.5	.6225	.7954	1.2572	.7826	.5
.6	.6239	.7983	1.2527	.7815	.4
.7	.6252	.8012	1.2482	.7804	.3
.8	.6266	.8040	1.2437	.7793	.2
.9	.6280	.8069	1.2393	.7782	51.1
39.0	0.6293	0.8098	1.2349	0.7771	**51.0**
.1	.6307	.8127	1.2305	.7760	50.9
.2	.6320	.8156	1.2261	.7749	.8
.3	.6334	.8185	1.2218	.7738	.7
.4	.6347	.8214	1.2174	.7727	.6
.5	.6361	.8243	1.2131	.7716	.5
.6	.6374	.8273	1.2088	.7705	.4
.7	.6388	.8302	1.2045	.7694	.3
.8	.6401	.8332	1.2002	.7683	.2
.9	.6414	.8361	1.1960	.7672	50.1
40.0	0.6428	0.8391	1.1918	0.7660	**50.0**
.1	.6441	.8421	1.1875	.7649	49.9
.2	.6455	.8451	1.1833	.7638	.8
.3	.6468	.8481	1.1792	.7627	.7
.4	.6481	.8511	1.1750	.7615	.6
40.5	0.6494	0.8541	1.1708	0.7604	**49.5**

Deg.	Sin	Tan	Cot	Cos	
40.5	0.6494	0.8541	1.1708	0.7604	**49.5**
.6	.6508	.8571	1.1667	.7593	.4
.7	.6521	.8601	1.1626	.7581	.3
.8	.6534	.8632	1.1585	.7570	.2
.9	.6547	.8662	1.1544	.7559	49.1
41.0	0.6561	0.8693	1.1504	0.7547	**49.0**
.1	.6574	.8724	1.1463	.7536	48.9
.2	.6587	.8754	1.1423	.7524	.8
.3	.6600	.8785	1.1383	.7513	.7
.4	.6613	.8816	1.1343	.7501	.6
.5	.6626	.8847	1.1303	.7490	.5
.6	.6639	.8878	1.1263	.7478	.4
.7	.6652	.8910	1.1224	.7466	.3
.8	.6665	.8941	1.1184	.7455	.2
.9	.6678	.8972	1.1145	.7443	48.1
42.0	0.6691	0.9004	1.1106	0.7431	**48.0**
.1	.6704	.9036	1.1067	.7420	47.9
.2	.6717	.9067	1.1028	.7408	.8
.3	.6730	.9099	1.0990	.7396	.7
.4	.6743	.9131	1.0951	.7385	.6
.5	.6756	.9163	1.0913	.7373	.5
.6	.6769	.9195	1.0875	.7361	.4
.7	.6782	.9228	1.0837	.7349	.3
.8	.6794	.9260	1.0799	.7337	.2
.9	.6807	.9293	1.0761	.7325	47.1
43.0	0.6820	0.9325	1.0724	0.7314	**47.0**
.1	.6833	.9358	1.0686	.7302	46.9
.2	.6845	.9391	1.0649	.7290	.8
.3	.6858	.9424	1.0612	.7278	.7
.4	.6871	.9457	1.0575	.7266	.6
.5	.6884	.9490	1.0538	.7254	.5
.6	.6896	.9523	1.0501	.7242	.4
.7	.6909	.9556	1.0464	.7230	.3
.8	.6921	.9590	1.0428	.7218	.2
.9	.6934	.9623	1.0392	.7206	46.1
44.0	0.6947	0.9657	1.0355	0.7193	**46.0**
.1	.6959	.9691	1.0319	.7181	45.9
.2	.6972	.9725	1.0283	.7169	**.8**
.3	.6984	.9759	1.0247	.7157	**.7**
.4	.6997	.9793	1.0212	.7145	.6
.5	.7009	.9827	1.0176	.7133	.5
.6	.7022	.9861	1.0141	.7120	.4
.7	.7034	.9896	1.0105	.7108	.3
.8	.7046	.9930	1.0070	.7096	.2
.9	.7059	.9965	1.0035	.7083	45.1
45.0	0.7071	1.0000	1.0000	0.7071	**45.0**
	Cos	Cot	Tan	Sin	Deg.

TABLE IV Trigonometric (Radians).

Rad.	Sin	Tan	Cot	Cos	Rad.	Sin	Tan	Cot	Cos
.00	.00000	.00000	∞	1.00000	**.50**	.47943	.54630	1.8305	.87758
.01	.01000	.01000	99.997	0.99995	.51	.48818	.55936	1.7878	.87274
.02	.02000	.02000	49.993	.99980	.52	.49688	.57256	1.7465	.86782
.03	.03000	.03001	33.323	.99955	.53	.50553	.58592	1.7067	.86281
.04	.03999	.04002	24.987	.99920	.54	.51414	.59943	1.6683	.85771
.05	.04998	.05004	19.983	.99875	.55	.52269	.61311	1.6310	.85252
.06	.05996	.06007	16.647	.99820	.56	.53119	.62695	1.5950	.84726
.07	.06994	.07011	14.262	.99755	.57	.53963	.64097	1.5601	.84190
.08	.07991	.08017	12.473	.99680	.58	.54802	.65517	1.5263	.83646
.09	.08988	.09024	11.081	.99595	.59	.55636	.66956	1.4935	.83094
.10	.09983	.10033	9.9666	.99500	**.60**	.56464	.68414	1.4617	.82534
.11	.10978	.11045	9.0542	.99396	.61	.57287	.69892	1.4308	.81965
.12	.11971	.12058	8.2933	.99281	.62	.58104	.71391	1.4007	.81388
.13	.12963	.13074	7.6489	.99156	.63	.58914	.72911	1.3715	.80803
.14	.13954	.14092	7.0961	.99022	.64	.59720	.74454	1.3431	.80210
.15	.14944	.15114	6.6166	.98877	.65	.60519	.76020	1.3154	.79608
.16	.15932	.16138	6.1966	.98723	.66	.61312	.77610	1.2885	.78999
.17	.16918	.17166	5.8256	.98558	.67	.62099	.79225	1.2622	.78382
.18	.17903	.18197	5.4954	.98384	.68	.62879	.80866	1.2366	.77757
.19	.18886	.19232	5.1997	.98200	.69	.63654	.82534	1.2116	.77125
.20	.19867	.20271	4.9332	**.98007**	**.70**	.64422	.84229	1.1872	.76484
.21	.20846	.21314	4.6917	.97803	.71	.65183	.85953	1.1634	.75836
.22	.21823	.22362	4.4719	.97590	.72	.65938	.87707	1.1402	.75181
.23	.22798	.23414	4.2709	.97367	.73	.66687	.89492	1.1174	.74517
.24	.23770	.24472	4.0864	.97134	.74	.67429	.91309	1.0952	.73847
.25	.24740	.25534	3.9163	.96891	.75	.68164	.93160	1.0734	.73169
.26	.25708	.26602	3.7591	.96639	.76	.68892	.95045	1.0521	.72484
.27	.26673	.27676	3.6133	.96377	.77	.69614	.96967	1.0313	.71791
.28	.27636	.28755	3.4776	.96106	.78	.70328	.98926	1.0109	.71091
.29	.28595	.29841	3.3511	.95824	.79	.71035	1.0092	.99084	.70385
.30	.29552	.30934	3.2327	.95534	**.80**	.71736	1.0296	.97121	.69671
.31	.30506	.32033	3.1218	.95233	.81	.72429	1.0505	.95197	.68950
.32	.31457	.33139	3.0176	.94924	.82	.73115	1.0717	.93309	.68222
.33	.32404	.34252	2.9195	.94604	.83	.73793	1.0934	.91455	.67488
.34	.33349	.35374	2.8270	.94275	.84	.74464	1.1156	.89635	.66746
.35	.34290	.36503	2.7395	.93937	.85	.75128	1.1383	.87848	.65998
.36	.35227	.37640	2.6567	.93590	.86	.75784	1.1616	.86091	.65244
.37	.36162	.38786	2.5782	.93233	.87	.76433	1.1853	.84365	.64483
.38	.37092	.39941	2.5037	.92866	.88	.77074	1.2097	.82668	.63715
.39	.38019	.41105	2.4328	.92491	.89	.77707	1.2346	.80998	.62941
.40	.38942	.42279	2.3652	.92106	**.90**	.78333	1.2602	.79355	.62161
.41	.39861	.43463	2.3008	.91712	.91	.78950	1.2864	.77738	.61375
.42	.40776	.44657	2.2393	.91309	.92	.79560	1.3133	.76146	.60582
.43	.41687	.45862	2.1804	.90897	.93	.80162	1.3409	.74578	.59783
.44	.42594	.47078	2.1241	.90475	.94	.80756	1.3692	.73034	.58979
.45	.43497	.48306	2.0702	.90045	.95	.81342	1.3984	.71511	.58168
.46	.44395	.49545	2.0184	.89605	.96	.81919	1.4284	.70010	.57352
.47	.45289	.50797	1.9686	.89157	.97	.82489	1.4592	.68531	.56530
.48	.46178	.52061	1.9208	.88699	.98	.83050	1.4910	.67071	.55702
.49	.47063	.53339	1.8748	.88233	.99	.83603	1.5237	.65631	.54869
.50	.47943	.54630	1.8305	.87758	**1.00**	.84147	1.5574	.64209	.54030
Rad.	Sin	Tan	Cot	Cos	Rad.	Sin	Tan	Cot	Cos

TABLE IV   Trigonometric (Radians) (*continued*).

Rad.	Sin	Tan	Cot	Cos	Rad.	Sin	Tan	Cot	Cos
1.00	.84147	1.5574	.64209	.54030	**1.50**	.99749	14.101	.07091	.07074
1.01	.84683	1.5922	.62806	.53186	1.51	.99815	16.428	.06087	.06076
1.02	.85211	1.6281	.61420	.52337	1.52	.99871	19.670	.05084	.05077
1.03	.85730	1.6652	.60051	.51482	1.53	.99917	24.498	.04082	.04079
1.04	.86240	1.7036	.58699	.50622	1.54	.99953	32.461	.03081	.03079
1.05	.86742	1.7433	.57362	.49757	1.55	.99978	48.078	.02080	.02079
1.06	.87236	1.7844	.56040	.48887	1.56	.99994	92.621	.01080	.01080
1.07	.87720	1.8270	.54734	.48012	1.57	1.00000	1255.8	.00080	.00080
1.08	.88196	1.8712	.53441	.47133	1.58	.99996	−108.65	− .00920	− .00920
1.09	.88663	1.9171	.52162	.46249	1.59	.99982	−52.067	− .01921	− .01920
1.10	.89121	1.9648	.50897	.45360	**1.60**	.99957	−34.233	− .02921	− .02920
1.11	.89570	2.0143	.49644	.44466	1.61	.99923	−25.495	− .03922	− .03919
1.12	.90010	2.0660	.48404	.43568	1.62	.99879	−20.307	− .04924	− .04918
1.13	.90441	2.1198	.47175	.42666	1.63	.99825	−16.871	− .05927	− .05917
1.14	.90863	2.1759	.45959	.41759	1.64	.99761	−14.427	− .06931	− .06915
1.15	.91276	2.2345	.44753	.40849	1.65	.99687	−12.599	− .07937	− .07912
1.16	.91680	2.2958	.43558	.39934	1.66	.99602	−11.181	− .08944	− .08909
1.17	.92075	2.3600	.42373	.39015	1.67	.99508	−10.047	− .09953	− .09904
1.18	.92461	2.4273	.41199	.38092	1.68	.99404	− 9.1208	− .10964	− .10899
1.19	.92837	2.4979	.40034	.37166	1.69	.99290	− 8.3492	− .11977	− .11892
1.20	.93204	2.5722	.38878	.36236	**1.70**	.99166	− 7.6966	− .12993	− .12884
1.21	.93562	2.6503	.37731	.35302	1.71	.99033	− 7.1373	− .14011	− .13875
1.22	.93910	2.7328	.36593	.34365	1.72	.98889	− 6.6524	− .15032	− .14865
1.23	.94249	2.8198	.35463	.33424	1.73	.98735	− 6.2281	− .16056	− .15853
1.24	.94578	2.9119	.34341	.32480	1.74	.98572	− 5.8535	− .17084	− .16840
1.25	.94898	3.0096	.33227	.31532	1.75	.98399	− 5.5204	− .18115	− .17825
1.26	.95209	3.1133	.32121	.30582	1.76	.98215	− 5.2221	− .19149	− .18808
1.27	.95510	3.2236	.31021	.29628	1.77	.98022	− 4.9534	− .20188	− .19789
1.28	.95802	3.3413	.29928	.28672	1.78	.97820	− 4.7101	− .21231	− .20768
1.29	.96084	3.4672	.28842	.27712	1.79	.97607	− 4.4887	− .22278	− .21745
1.30	.96356	3.6021	.27762	.26750	**1.80**	.97385	− 4.2863	− .23330	− .22720
1.31	.96618	3.7471	.26687	.25785	1.81	.97153	− 4.1005	− .24387	− .23693
1.32	.96872	3.9033	.25619	.24818	1.82	.96911	− 3.9294	− .25449	− .24663
1.33	.97115	4.0723	.24556	.23848	1.83	.96659	− 3.7712	− .26517	− .25631
1.34	.97348	4.2556	.23498	.22875	1.84	.96398	− 3.6245	− .27590	− .26596
1.35	.97572	4.4552	.22446	.21901	1.85	.96128	− 3.4881	− .28669	− .27559
1.36	.97786	4.6734	.21398	.20924	1.86	.95847	− 3.3608	− .29755	− .28519
1.37	.97991	4.9131	.20354	.19945	1.87	.95557	− 2.2419	− .30846	− .29476
1.38	.98185	5.1774	.19315	.18964	1.88	.95258	− 3.1304	− .31945	− .30430
1.39	.98370	5.4707	.18279	.17981	1.89	.94949	− 3.0257	− .33051	− .31381
1.40	.98545	5.7979	.17248	.16997	**1.90**	.94630	− 2.9271	− .34164	− .32329
1.41	.98710	6.1654	.16220	.16010	1.91	.94302	− 2.8341	− .35284	− .33274
1.42	.98865	6.5811	.15195	.15023	1.92	.93965	− 2.7463	− .36413	− .34215
1.43	.99010	7.0555	.14173	.14033	1.93	.93618	− 2.6632	− .37549	− .35153
1.44	.99146	7.6018	.13155	.13042	1.94	.93262	− 2.5843	− .38695	− .36087
1.45	.99271	8.2381	.12139	.12050	1.95	.92896	− 2.5095	− .39849	− .37018
1.46	.99387	8.9886	.11125	.11057	1.96	.92521	− 2.4383	− .41012	− .37945
1.47	.99492	9.8874	.10114	.10063	1.97	.92137	− 2.3705	− .42185	− .38868
1.48	.99588	10.983	.09105	.09067	1.98	.91744	− 2.3058	− .43368	− .39788
1.49	.99674	12.350	.08097	.08071	1.99	.91341	− 2.2441	− .44562	− .40703
1.50	.99749	14.101	.07091	.07074	**2.00**	.90930	− 2.1850	− .45766	− .41615
Rad.	Sin	Tan	Cot	Cos	Rad.	Sin	Tan	Cot	Cos

INDEX

ANGLE

An *angle* is formed by rotating a ray about its endpoint from some initial position to some terminal position. The measure of an angle is the amount of rotation. If the angle's vertex is at the origin of a Cartesian coordinate system and its initial side is along the positive x-axis, the angle is said to be in *standard position*.

ARC LENGTH

The *arc length*, s, cut by a central angle θ, measured in radians, of a circle of radius r is given by $s = r\theta$.

TRIGONOMETRIC FUNCTIONS

Let θ be an angle in standard position with a point $P(x, y)$ on the terminal side a distance of $r = \sqrt{x^2 + y^2}$ from the origin $(r \neq 0)$. Then we define the *trigonometric functions* as follows:

$$\cos \theta = \frac{x}{r} \qquad \tan \theta = \frac{y}{x}, \ x \neq 0 \qquad \csc \theta = \frac{r}{y}, \ y \neq 0$$

$$\sin \theta = \frac{y}{r} \qquad \sec \theta = \frac{r}{x}, \ x \neq 0 \qquad \cot \theta = \frac{x}{y}, \ y \neq 0$$

EXACT VALUES

angle θ / functions	0	$\frac{\pi}{6}$	$\frac{\pi}{4}$	$\frac{\pi}{3}$	$\frac{\pi}{2}$	π	$\frac{3\pi}{2}$
$\cos \theta$	1	$\frac{\sqrt{3}}{2}$	$\frac{\sqrt{2}}{2}$	$\frac{1}{2}$	0	-1	0
$\sin \theta$	0	$\frac{1}{2}$	$\frac{\sqrt{2}}{2}$	$\frac{\sqrt{3}}{2}$	1	0	-1
$\tan \theta$	0	$\frac{\sqrt{3}}{3}$	1	$\sqrt{3}$	undef.	0	undef.
$\sec \theta$	1	$\frac{2}{\sqrt{3}}$	$\frac{2}{\sqrt{2}}$	2	undef.	-1	undef.
$\csc \theta$	undef.	2	$\frac{2}{\sqrt{2}}$	$\frac{2}{\sqrt{3}}$	1	undef.	-1
$\cot \theta$	undef.	$\frac{3}{\sqrt{3}}$	1	$\frac{\sqrt{3}}{3}$	0	undef.	0

INVERSE TRIGONOMETRIC FUNCTIONS

Inverse Function	Domain	Range
$y = \text{Arccos } x$ or $y = \text{Cos}^{-1}x$	$-1 \leq x \leq 1$	$0 \leq y \leq \pi$
$y = \text{Arcsin } x$ or $y = \text{Sin}^{-1}x$	$-1 \leq x \leq 1$	$-\pi/2 \leq y \leq \pi/2$
$y = \text{Arctan } x$ or $y = \text{Tan}^{-1}x$	all reals	$-\pi/2 < y < \pi/2$
$y = \text{Arccot } x$ or $y = \text{Cot}^{-1}x$	all reals	$0 < y < \pi$